Supply chain integration in the building industry

Stellingen bij het proefschrift

Ketenintegratie in de bouw: de opkomst van geïntegreerde en projectongebonden strategieën in een gefragmenteerde en projectgedreven sector

Door Ruben Vrijhoef

1. De bouw zal altijd een projectmatig werkende sector blijven (dit proefschrift).
2. Het vergelijken van de bouw met andere sectoren kan alleen worden uitgelegd als interesse in andere sectoren of als medelijden met de bouw (dit proefschrift).
3. De integratie van de aanbodketen in de bouw zal haar doel nooit bereiken zonder de integratie van de vraagketen (dit proefschrift).
4. Het oplijnen van de bijdragen van bedrijven en personen aan de keten verbetert de gezamenlijke prestatie alsmede de individuele prestatie (dit proefschrift).
5. In de bouw is sprake van een fundamentele mismatch tussen de vraag die dynamisch van aard is en het aanbod van producten die statisch van aard zijn.
6. De opdeling van het parlement in links en rechts doet geen recht aan de complexe en dynamische maatschappij die het vertegenwoordigt.
7. De teruggang van de wereldwijde economie is een vooruitgang voor de wereld als geheel.
8. Wie een kuil graaft voor een ander is een altruïst.
9. Het resultaat van een promotieonderzoek ligt veeleer in de promovendus dan in het proefschrift.
10. De publicatiedrift aan universiteiten wordt feitelijk ingegeven door oudhollandse logica: Wie schrijft, die blijft.

Deze stellingen worden opponeerbaar en verdedigbaar geacht en zijn als zodanig goedgekeurd door de promotoren prof.dr.ir. H.A.J. de Ridder, prof.ir. H. de Jonge en prof.dr.ir. B.A.G. Bossink.

Propositions to the thesis

Supply chain integration in the building industry: the emergence of integrated and repetitive strategies in a fragmented and project-driven industry
By Ruben Vrijhoef

1. The building industry will always be a project-driven sector of industry (this thesis).
2. Comparing the building industry to other sectors of industry can only be explained either by interest in those other sectors or sympathy for the building industry (this thesis).
3. Supply chain integration in the building industry will never be achieved without demand chain integration (this thesis).
4. Aligning the contributions of firms and individuals to the supply chain improves the joint performance as well as the individual performances (this thesis).
5. In the building industry there is a fundamental mismatch between the demand that is typically dynamic and the supply of products that are typically static.
6. The division of the parliament into a left and a right wing does not do justice to the complex and dynamic society it represents.
7. The decline of the worldwide economy implies an improvement of the world as a whole.
8. One who digs a hole for another is an altruist.
9. The result of a doctoral study lies rather within the researcher than in the thesis.
10. The urge to publish at universities is actually inspired by old Dutch logic: One who writes, will stay.

These propositions are regarded as opposable and defendable, and have been approved as such by the supervisors prof.dr.ir. H.A.J. de Ridder, prof.ir. H. de Jonge and prof.dr.ir. B.A.G. Bossink.

Supply chain integration in the building industry

The emergence of integrated and repetitive strategies in a fragmented and project-driven industry

Proefschrift

ter verkrijging van de graad van doctor
aan de Technische Universiteit Delft,
op gezag van de Rector Magnificus
prof. ir. K.C.A.M Luyben,
voorzitter van het College voor Promoties,
in het openbaar te verdedigen op
25 november 2011 om 10:00 uur

door Ruben VRIJHOEF

civiel ingenieur
geboren 26 december 1972 te Wormerveer.

Dit proefschrift is goedgekeurd door de promotoren:
Prof.dr.ir. H.A.J. de Ridder
Prof.ir. H. de Jonge
Prof.dr.ir. B.A.G. Bossink

Samenstelling promotiecommissie:
Rector Magnificus, voorzitter
Prof.dr.ir. H.A.J. de Ridder, Technische Universiteit Delft, promotor
Prof.ir. H. de Jonge, Technische Universiteit Delft, promotor
Prof.dr.ir. B.A.G. Bossink, Vrije Universiteit Amsterdam, promotor
Prof.dr. P.S. Barrett, University of Salford
Prof.dr. W.P. Hughes, University of Reading
Prof.dr. J.A.A. van der Veen, Universiteit van Amsterdam
Prof.mr.dr.ir. S.C. Santema, Technische Universiteit Delft
Prof.dr.ir. J.W.F. Wamelink, Technische Universiteit Delft

ISBN 978-1-60750-962-2 (print)

Keywords: building, manufacturing, supply chain, integration, comparison.

Published and distributed by IOS Press under the imprint Delft University Press

Publisher
IOS Press
Nieuwe Hemweg 6b
1013 BG Amsterdam
The Netherlands
tel: +31-20-688 3355
fax: +31-20-687 0019
email: info@iospress.nl
www.iospress.nl

PRINTED IN THE NETHERLANDS

Preface

The building industry is one of the oldest industries around. The ways of working within the industry have developed at their own pace over the years. The industry has been applauded for furnishing great works, providing populations with accommodations and infrastructures, and substantially contributing to economic growth and prosperity worldwide. However, the industry has also been criticised extensively for being wasteful, not innovative and unproductive. In those criticisms, reference is often made to other sectors of industry, particularly manufacturing, that functions and produces its products in more effective and efficient ways than the building industry does apparently. One of the major differences observed between building and manufacturing, among other things, has been the organisation and coordination of the supply chain. Such differences call for a careful consideration of the possibilities and impossibilities of transferring supply chain approaches from manufacturing to building.

The organisation and coordination of supply chains have been observed and conceptualised in different ways in building and manufacturing based on the specifically characteristic differences between those sectors. In building, the organisational approaches of supply chains have particularly been influenced by the one-off, temporal nature of projects; the large number of firms involved in the definition, design, manufacture and assembly of built objects involving many relatively small firms; the dispersed power and governance regimes; and the initiating role of clients. These characteristics and other specific aspects of building have influenced how firms in the building industry operate, how they manage their inter-firm relationships, and in essence, how the industry as a whole is organised. In order to address improvements of this situation, managerial and organisational arrangements between firms in the building supply chain need to be reconsidered.

A more integrated approach to the supply chain has been suggested as a solution to the many problems and deficiencies existing in building. On the other hand, the restrictions on increasing the level of integration in building also need to be taken into account. The approach's underlying principle would be that the supply chain that is delivering a single product should not be fragmented nor consist of disconnected functions. Instead, supply chain integration would lead to a more stable and repetitive production environment, similar to what is common in manufacturing. The premise here is that the building supply chain would function better when approached and reconceptualised as a single entity, an extended enterprise. In a way, the deeper issue here is whether the building industry could or should develop itself towards the standards and practices of a more integrated and repetitively operating industry, such as manufacturing.

This thesis represents an exploration of that idea, and marks the end of a longer journey in that direction that has been ongoing for some years. A selection of parts of previous writings produced during this journey has become an integrated part of this thesis. The aspiration of the thesis is that it will contribute to the theoretical and practical development of the concept of supply chain integration and the positive effects it may yield for the building industry.

Ruben Vrijhoef

Delft
November 2011

Acknowledgements

Doing a doctoral study and writing up the thesis is a lonely task. It is actually meant as a test for the researcher, to prove whether he or she is competent to perform such a task individually. However, it cannot really be done without the help of many. In fact, it is thanks to the support of many that this thesis is now finally lying before you. First, my thanks go to my supervisors, Hennes de Ridder, Hans de Jonge and Bart Bossink, for their continual support, criticism and patience. I want to thank Bart in particular for his help in getting me through the final process of writing up the thesis. I could not have done it without him. In addition, I want to thank Hans Wamelink for being so generous in allowing me to take some time off to write up the thesis. I want to thank Jelle Koolwijk for taking over my activities at the university and the centre during my absence. Thanks also to other colleagues at the university, clients and industry partners for their expressions of interest in my work and the thesis. In particular, I want to thank the PSIBouw innovation programme, Delft University of Technology and the Centre for Process Innovation in Building and Construction for their financial support of this thesis.

In the wider community, I want to thank my many colleagues around the globe for our inspiring debates at the many occasions and conferences where we met and discussed the subject of the thesis. Within this community, I want to thank a few distinguished people in particular. Thanks go to my first academic mentor and supervisor, Lauri Koskela, for introducing me in 1997 to this intriguing realm of research while I was finishing my master thesis in Finland, and selling me on the idea of embarking on a doctoral study in the first place. Parts of this thesis are based on previous papers I wrote that have now shown to be part of the longer journey of writing this thesis bit by bit during the last couple of years. In a few cases, the papers were co-written with colleagues around the globe, notably Kerry London, William O'Brien, Carlos Formoso, Hans Voordijk and Lauri Koskela. Thanks to you all for the collaborative writing and being able to reuse and adapt a few parts of the texts, particularly in the first part of the thesis. With respect to the empirical part of the thesis, particularly the building case studies, I want to thank all respondents and informants involved for contributing their input and time so generously. I would like to thank Roberta Hardy for her valuable corrections of the English, and Corien Smit for doing a wonderful job on the design and the layout of the thesis.

To conclude, my thanks go to the smaller community of my friends, relatives and family. I want to thank you all equally for your interest, support and the distraction from this work I needed every now and then. There are four people I want to thank especially: my mother-in-law, Wil, for helping me out with the transcripts of the interviews, and last and most especially, my thanks go to my dearest Marianne, Pepijn and Meinou for your endless confidence, patience and love. This thesis is dedicated to you.

Table of Contents

Preface v

Acknowledgements vii

1 Introduction 1

1.1 Background: understanding the building industry as a project-driven industry 1
1.2 Research problem: fragmentation and lack of repetition in the building supply chain ... 4
1.3 Rationale: the concept of supply chain integration as a development path for the building supply chain 7
1.4 Research objective: specifying a concept of supply chain integration for the building industry 11
1.5 Research questions: achieving the objective of a concept of supply chain integration for the building industry 13
1.6 Limitations and justification: scope and relevance 14
1.7 Structure of the thesis: parts and chapters 15
1.8 Reflection on key issues and contributions of this chapter to the thesis 17

2 Research design 19

2.1 Research approach: contributing to theory and practice 19
2.1.1 This research viewed as organisational qualitative research 19
2.1.2 This research viewed as knowledge production 21
2.1.3 This research viewed as theory building 22

2.2 Research strategy: theory building from case study research 23
2.3 Structure of the research: research stages and activities 26
2.3.1 Part I: Theoretical basis 26
2.3.2 Part II: Empirical analysis 26
2.3.3 Part III: Synthesis 28
2.4 Justification of the research approach 30
2.4.1 Validity 30
2.4.2 Reliability 31
2.4.3 Generalisability 31
2.5 Reflection on key issues and contributions of this chapter to the thesis 31

PART I: THEORETICAL BASIS 35

3 Framing the concept of supply chain integration in building 37

3.1 The idea to learn and transfer supply chain concepts from other industries to the building industry 39
3.1.1 General lessons from manufacturing as potential solutions for building 39
3.1.2 Differences between industries as a potential hindrance transferring concepts 40
3.2 Previous thoughts pointing towards supply chain integration in building 41
3.2.1 Quasi-firm and extended enterprise 41
3.2.2 Supply chain management 42
3.2.3 System integrator 43
3.3 Contributing topics to supply chain integration in the building industry 44
3.3.1 Integrated contracts and strategic procurement 45
3.3.2 Alliances and partnering 46
3.3.3 Lean and agile construction 47
3.3.4 Co-makership 47
3.3.5 Just-in-time logistics 48
3.3.6 Modularisation, standardisation and postponement 48
3.3.7 Prefabrication, offsite construction and industrialisation 49
3.4 Reflection on key issues and contributions of this chapter to the thesis 49

4 Theoretical framework 53

4.1 Explaining the need for a multiple theoretical view on the supply chain and the selection of the four perspectives 53
4.2 Introducing the four theoretical perspectives on the supply chain 57

4.2.1 Economic perspective: transaction cost economics 57
4.2.2 Production perspective: production management 59
4.2.3 Organisational perspective: inter-firm networks 60
4.2.4 Social perspective: communication and commitment 61
4.3 Applying the four theoretical perspectives to the building supply chain 61
4.3.1 Economic perspective: transaction cost economics 61
4.3.2 Production perspective: production management 63
4.3.3 Organisational perspective: inter-firm networks 63
4.3.4 Social perspective: communication and commitment 64
4.4 Synthesising the theoretical framework 65
4.5 Reflection on key issues and contributions of this chapter to the thesis 66

5 Analytical model for supply chain integration 69

5.1 Operationalising the theoretical framework into factors of analysis 69
5.1.1 Economic governance of the integrated supply chain 73
5.1.2 Production management of the integrated supply chain 74
5.1.3 Inter-organisational governance of the integrated supply chain 75
5.1.4 Social governance of the integrated supply chain 77
5.2 Identifying four development levels of supply chain integration 80
5.3 Relating the factors of analysis to the four development levels 81
5.4 Reflection on key issues and contributions of this chapter to the thesis 84

PART II: EMPIRICAL ANALYSIS 87

6 Supply chain integration practices in manufacturing 89

6.1 Implications of conceptual investigations across industries 89
6.2 Explanation of the selection of manufacturing industries 91
6.2.1 Industry typology 91
6.2.2 Supply chain typology 92
6.2.3 Industries selected for this thesis 95
6.3 Supply chain integration practices applied in the automotive industry 96
6.3.1 Example of consumer cars: Toyota 98
6.4 Supply chain integration practices applied in the aerospace industry 101
6.4.1 Example of aircrafts: Airbus 103
6.5 Supply chain integration practices applied in the computer industry 106
6.5.1 Example of personal computers: Dell 107
6.6 Supply chain integration practices applied in the electronics industry 110
6.6.1 Example of semiconductor machines: ASML 112

6.7 Supply chain integration practices applied in the clothing and textile industry 115
6.7.1 Example of fashion: Zara 117
6.8 Supply chain integration practices applied in the food and grocery sector 121
6.8.1 Example of groceries: Wal-Mart 123
6.9 Reflection on key issues and contributions of this chapter to the thesis 128

7 Supply chain integration practices in the building industry 131

7.1 Outline of the building case studies 131
7.1.1 Cases selected for this thesis 131
7.1.2 Organisation of case descriptions 134
7.2 Supply chain integration practices applied by clients 134
7.2.1 Case of a housing corporation in operational cooperation with builders (Client 1) 134
7.2.2 Case of a housing corporation in legal partnership with builders (Client 2) 140
7.3 Supply chain integration practices applied by developers 145
7.3.1 Case of a full-service project developer (Developer) 145
7.4 Supply chain integration practices applied by designers 149
7.4.1 Case of a developing and contracting architect (Designer 1) 149
7.4.2 Case of an architectural engineer (Designer 2) 153
7.5 Supply chain integration practices applied by builders 156
7.5.1 Case of a catalogue builder (Builder 1) 156
7.5.2 Case of an ICT-driven builder (Builder 2) 160
7.5.3 Case of a lean-driven builder (Builder 3) 164
7.6 Supply chain integration practices applied by suppliers 168
7.6.1 Case of three suppliers integrating products and site installation (Suppliers) 168
7.7 Reflection on key issues and contributions of this chapter to the thesis 173

PART III: SYNTHESIS 177

8 Case study analyses 179

8.1 Analyses of the building case studies 179
8.1.1 Within-case analyses of the building cases 179
8.1.2 Cross-case analysis between the building cases 190
8.2 Analyses of the manufacturing comparator studies 192
8.2.1 Within-case analyses of the manufacturing comparators 192

8.2.2 Cross-case analysis between the manufacturing comparators 196
8.3 Comparative analysis between building and manufacturing 200
8.4 Reflection on key issues and contributions of this chapter to the thesis 201

9 Conclusions 207

9.1 Theory building: confronting theory with practice 207
9.1.1 Identifying the relationships supporting the hypotheses 209
9.1.2 Relating and clustering the theoretical factors 220
9.1.3 Relating and clustering the goals 223
9.1.4 Shaping the hypotheses 228
9.2 Answers to research questions 232
9.3 Contributions to theory and method 234
9.4 Implications for building practice 235
9.5 Directions for further research 237

References 243

Appendices 281

Case Study Protocol 281
Themes retrieved from coding 284
Data gathering in the manufacturing cases 286
Data gathering in the building cases 288
Glossary of Abbrevations and Terms 290
List of Figures 297
List of Tables 299
Summary 305
Samenvatting (summary in Dutch) 311
Curriculum Vitae 317
Publications 319

1 | Introduction

The building industry can be typified as a project-based industry with specific characteristics such as location-bound design, one-off production, changing production coalitions per project, outdoor and environmental circumstances, multiple clients and multiple suppliers involved in a single project. In comparison to many other industries, the production environment in building is relatively complex and unstable. This generally leads to negative effects, such as low levels of effectiveness and efficiency, low rates of innovation and impediments to knowledge sharing and learning. The performance level of the building sector is considered to be lower and lagging behind other industrial sectors. The introduction of more integrated and multi-project ways of working and collaboration such as those seen in other industrial sectors would seem logical and beneficial. One pathway towards these kinds of solutions is provided by the *concept of supply chain integration*. In manufacturing, supply chains have typically been integrated by focal companies, linking and synchronising suppliers' processes to their own business processes through applications of supply chain integration. This thesis represents a quest to construct a concept for supply chain integration in the building industry.

1.1 Background: understanding the building industry as a project-driven industry

In previous research as well as in practice, the building industry has been criticised for its supposed low level of performance and backwardness in many respects

(Woudhuysen & Abley 2004). The causes of the problematic character of building apparently lie in the very *nature of building*, and have been sought at the level of the product, the production in projects, and the industry as a whole. The nature of building has been blamed for contributing to waste and value loss, and it has been claimed that it is necessary to transform this or at least to mitigate the impact on the level of production (Koskela 2000). However, to achieve 'full resolution', it often seems that a particular characteristic has to be mastered at multiple levels of the production system. The industry's characteristics on the production level are related to characteristics on both the product and industry levels (Figure 1). The three levels of characteristics reinforce each other in a complex interaction, which contributes to the difficulty of reducing the problems of building and thus also to the persistence of the *problematic character* of building (Koskela 2000).

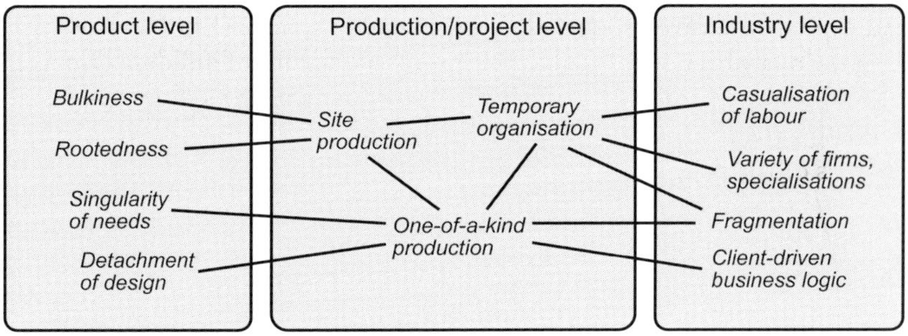

Figure 1: Three characteristics of building on product, production and industry level. Adapted from (Vrijhoef & Koskela 2005b)

The *basic characteristics* of built objects have been viewed as causing limitations to technology and problems in the management of building projects (Nam & Tatum 1988). In this context, various features have been mentioned, such as immobility, complexity, long product life cycle, capital intensity and impact on the environment. In addition, built objects are often unique with additional specific features, and they are built in a specific institutional and socio-economic context. At the industry level, there are high levels of *fragmentation*, a wide variety of firms of different specialisation and size, and high levels of casualisation of labour. In some cases it has even been questioned whether building can be regarded as an actual industry (Groák 1994), or more properly as a 'loosely coupled system' of projects (Dubois & Gadde 2002). Paradoxically, however, fragmentation of the industry must not be seen as strictly problematic. The involvement of many different specialised firms in projects does not necessarily cause low levels of efficiency. On the contrary, it has been claimed that this could just as well increase the efficiency of resource allocation and speed of information exchange between parties (Pryke 2002).

Still, the product, process and industry characteristics of building do have an impact on the *production situation* and the way in which building projects are organised. Building projects have been described as *coalitions of firms*; i.e. 'a number of independent firms coming together for the purpose of undertaking a single building project and that coalition of firms having to work as if it were a single firm, for the purposes of the project' (Winch 1989). The parties involved in building projects have been interpreted as 'organisational units joining and operating together as a single production organisation when it is advantageous' (Harland et al. 1999); a 'temporary multiple organisation' (Cherns & Bryant 1984); or a 'quasi-firm' (Eccles 1981). The production system has been regarded as 'capability-oriented production' (Wortmann 1992), and is always locally bound and thus dependent on physical factors such as soil and weather conditions.

The *organisation of production* and the supply chains is strongly aimed at the convergence of logistics to a particular site, and delivery of the one-off, customised and capital-intensive product to a single end customer (Lin & Shaw 1998). This has previously been identified as the 'prototype nature' of building (Koskela 2000), reflected most characteristically by the predominant one-off approach in discrete building projects, i.e. 'unique-product production' (Drucker 1963). These characteristics of the production organisation in the building industry are not unique as such and can also be found in other sectors of industry, but it is the specific *combination of characteristics*, which apparently makes the building situation unique. This implies that *concepts from other industries* could be applicable to the building industry if the combined characteristics of building and their causal relationships are addressed (Figure 2). The characteristics of building could be overcome or resolved in practice by *adapting* those concepts and *translating* them into a building context.

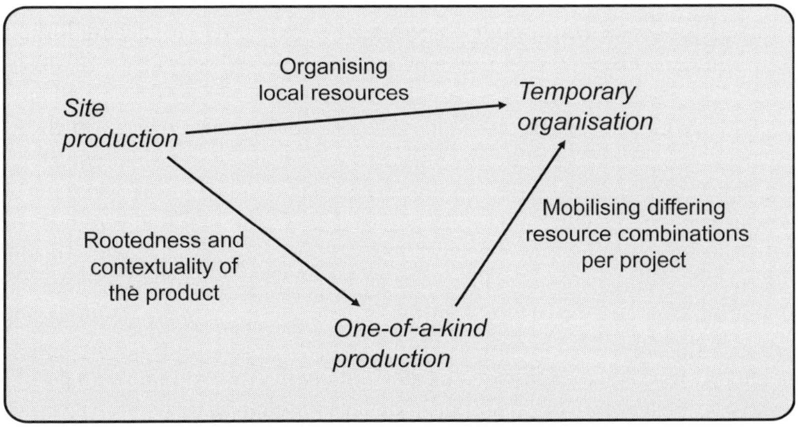

Figure 2: Characteristics of building and their causal relations (Vrijhoef & Koskela 2005b)

1.2 Research problem: fragmentation and lack of repetition in the building supply chain

Because of its project-based approach, the production system of the building industry is highly flexible and fragmented, consisting of many different subsectors (residential, commercial, etc.), many different disciplines (developers, builders, engineers, architects, etc.), and a wide spectrum of firm sizes with a relatively high share of SMEs. This has led to relatively high levels of *fragmentation* of the building supply chain from supplier to end user. Building projects are usually initiated by a client organisation, such as a housing corporation. On both the demand and the supply side, many parties play a role, including a large number of stakeholders on the demand side as well as a large number of co-producers on the supply side (Figure 3).

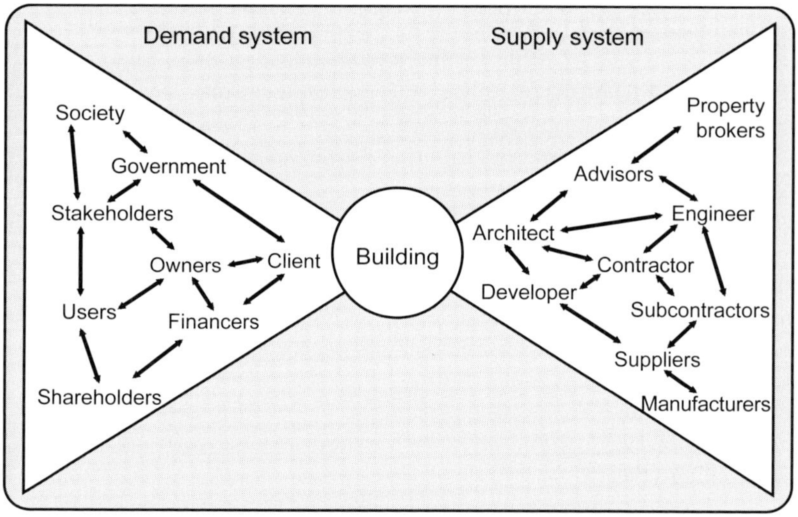

Figure 3: Schematic representation of the demand and supply system around a built object (Vrijhoef & De Ridder 2005)

Firms in the building industry work together in constantly changing coalitions on different building projects (O'Brien et al. 1995). This is particularly true in a traditional building setting, where multiple bilateral contracts are negotiated between individual parties who are involved in a temporary coalition until the completion of the project. As a result, the constructed product is seldom predefined, but instead the delivery of built products can be typified as assemble-to-order, make-to-order, design-to-order or even concept-to-order (Luhtala et al. 1994, Winch 2003) (Figure 4). This makes building essentially a demand-driven process that is mostly initiated by client organisations or investors. Design is often treated as an independent activity in the building process, disconnected from production. The production is very much influenced by craftsmanship, involving different crafts by specialist firms.

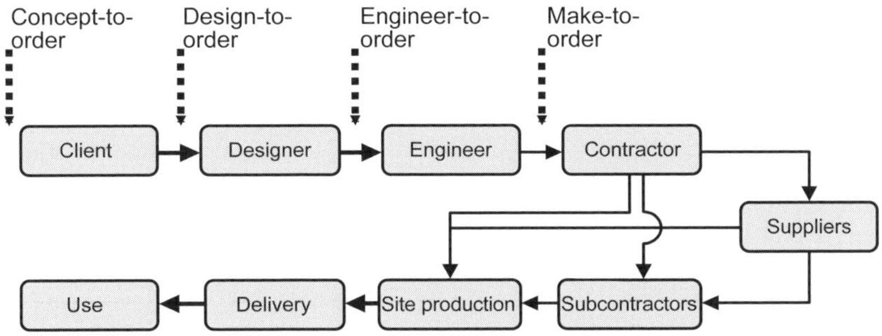

Figure 4: Different delivery types in the building supply chain form make-to-order to concept-to-order (Bossink & Vrijhoef 2009)

In addition to the high level of fragmentation, the level of *repetition* in building is low compared to many other industries. The production system types of different industries vary from (one-off) designing to (repetitive) making (Figure 5). In this view, making refers to manufacturing of assemble-to-order, make-to-order, or make-to-stock types of products. 'Treating building as a type of manufacturing obviously neglects the importance of the design aspect of building, and arguably subordinates value generation to waste reduction, which inverts their proper relationship'; however, 'certain aspects of building could move into the realm of repetitive making' (Ballard 2005).

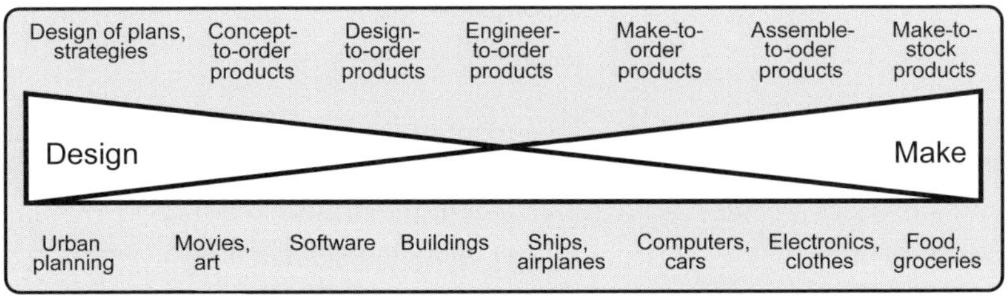

Figure 5: Production system types varying from (one-off) designing to (repetitive) making. The production system of building is deemed to be engineer-to-order (Adapted from Ballard 2005)

The low levels of repetition increase the *unpredictability* of the flow of work, and main contractors therefore generally outsource most of their turnover and purchase high amounts of labour and material. As a consequence, main contractors have become increasingly reliant on other firms in the supply chain, notably suppliers and subcontractors. In addition, within a contractor's organisation, the management function is typically disconnected from the production function on site. 'We have virtually two separate organisations: one for the management function and one for getting the work done. The two organisations do not coordinate their work, and they are characterised by different goals and viewpoints' (Applebaum 1982). This *separation of functions* decreases the degree of control in the supply chain, and increases the degree of fragmentation. This is worsened by the large number of SMEs and self-employed workers in the Dutch building industry (Figure 6), a result of the increasing casualisation among workers in most economic sectors, including the building industry (EIB 2009).

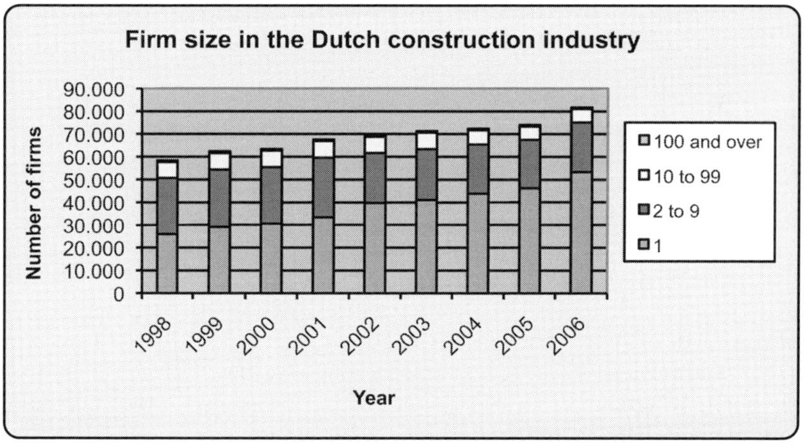

Figure 6: Changes in firm size in the Dutch building industry 1998-2006. Data provided by Bouwend Nederland on www.bouwendnederland.nl

Paradoxically, the high levels of fragmentation and the low levels of repetition in the building supply chain have led to many problems and deficiencies in the production system, as well as adaptability and progress in the industry (Pryke 2002). In the building supply chain, the two phenomena have led to typical *problems* of lack of control and decreasing performance, which tend to reinforce each other throughout the supply chain because of causal relationships within the supply chain (Figure 7). For instance, difficulties finding out the client's demands and design changes often lead to inaccurate data being handed over to contractors and suppliers, incorrect building requiring amendments on site, leading to delayed delivery of the building to the client and ultimately, delayed occupation of the building by the users.

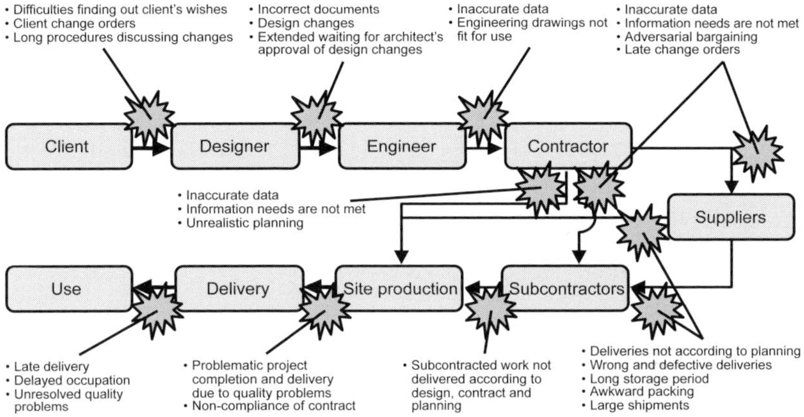

Figure 7: Problems progressing through the building supply chain (Vrijhoef 1998)

1.3 Rationale: the concept of supply chain integration as a development path for the building supply chain

The concept of supply chain integration has been described as a *development path* towards lower levels of fragmentation and higher levels of repetition in the supply chain (Cagliano et al. 2006, Campbell & Sankaran 2005). For the building supply chain, this would imply the establishment of a more stable production environment, for instance by a multi-project approach, and installing integrated process formats replacing the existing disintegrated and one-off production strategies. As advocated by many, this should bring improvement, but the idea has also been criticised as laden with rhetoric and been termed an 'elusive goal' (Briscoe & Dainty 2005, Fawcett & Magnan 2002). Notwithstanding those qualifications, supply chain integration as a concept aids in looking across the entire supply chain rather than just at the next entity or level. It aims to increase transparency and alignment of the supply chain's coordination and configuration, regardless of functional or corporate boundaries (Cooper & Ellram 1993). In other words, supply chain integration aims to shift the supply chain to a single 'extended enterprise' or one 'virtual organisation' (Cagliano et al. 2005, Shekhar 2006). In general terms, such representations of supply chain integration would imply a development path for the building supply chain towards improvement of effectiveness and efficiency.

As a result, supply chain integration has been considered a *solution* to many of the problems and to the underperformance of the building supply chain, for example, by establishing long-term multi-project partnering arrangements between clients and firms (Bresnen & Marshall 2000a). On the other hand, it has been questioned whether the large numbers and different types of firms reflecting the high levels of fragmentation and the ever-changing constellations of firms in building projects may hinder supply chain integration (Briscoe & Dainty 2005). Completely integrated supply chains involving all firms in continual repetitive works would probably be unlikely to be achieved in building practice. Nevertheless, supply chain integration should at least be possible to some extent by following the basic principles of supply chain integration, such as those utilised in manufacturing, and when developing corresponding collaborative methods, new approaches to adding value, and new capabilities, not the least of which being systems integration (Brady et al. 2005).

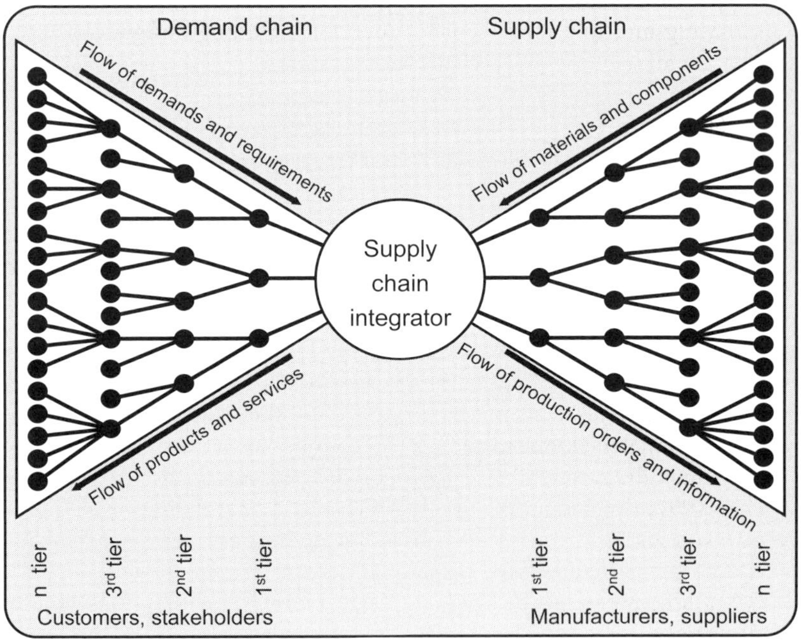

Figure 8: The idea of the supply chain integrator in building tying together and coordinating different flows through the supply chain. Adapted from (Segerstedt & Olofsson 2010, Vrijhoef & De Ridder 2005)

This may generally be true, but there are significant *barriers* to supply chain integration by smaller subcontractors and suppliers for instance. To a great extent, 'there remains a general mistrust within the SME companies that make up the building supply chain and a general lack of belief that there are mutual benefits in supply chain integration practices' (Dainty et al. 2001a). It is suggested that leading clients should take responsibility for engendering the necessary attitudinal change throughout the supplier networks if further performance is to be realised within the sector. Despite the potential barriers, previous work mostly points to the need for more *alignment* and more structured ways of organising and managing the building supply chain. Many have put forward the need for a *fundamental reconceptualisation* of existing building practice by applying supply chain principles from manufacturing and theory to the specific context of building. Such principles include the introduction of the role of the *supply chain integrator* in the supply chain, i.e. the focal firm tying together and coordinating all flows through the supply chain as it were an *extended enterprise* (Figure 8). In a building context, this could be a leading client such as a housing corporation entering into supply chain arrangements with contractors, suppliers, etc. From the supply side this could be a leading contractor, for instance, incorporating and coordinating multiple supply chain functions including design, construction and materials manufacture.

Supply chain integration compared to related supply chain concepts

Various *supply chain concepts* have emerged in parallel the last two or three decades in generic theory and manufacturing practice. These concepts are highly related and show much conceptual overlap This has led to much ambiguity between the definitions of the different concepts. Generally all supply chain concepts have originated from logistics and materials management (Christopher 1992). Gradually the concepts have evolved towards broader approaches to the supply chain, including additional aspects such as marketing and supplier involvement in product development (Christopher 1997, Cooper et al. 1997). Along this development *supply chain management* evolved from merely focussing on inventory planning and logistics management towards more comprehensive outsourcing strategies for instance including economic issues and risk sharing with suppliers (Bechtel & Jayaram 1997, Burgess et al. 2006, Chandra & Kumar 2000, Jones & Riley 1985, Williamson 2008).

In addition to supply chain management, *supply chain collaboration* and *cooperation* often include the establishment of collaborative systems such as joint Efficient Customer Response and Vendor Managed Inventory systems, particularly in fast consumer goods and retail sectors (Holweg et al. 2005, Soosay et al. 2008). Cooperative supply and production approaches may also exist among firms, such as district or distributed manufacture and supply solutions together with co-producers, suppliers and logistics service providers for instance (Albino et al. 2007). Further to supply chain collaboration and cooperation, *supply chain alliances* and *partnerships* have fostered more equal relations between supply chain firms viewed as partners. Often such alliances and partnerships reside on a strategic or tactical level between the firms involved, for instance including integrated arrangements to finance and risk sharing, but not necessarily collaborative management of activities on an operational level (Ellram & Cooper 1990, Lambert et al. 1996, Persson & Virum 2001).

In addition to the above supply chain concepts, *supply chain integration* appears to refer to higher levels of integration, notably integration of shared information and integrated management of supply chain activities and operations including product development, materials manufacture and product assembly (Hewitt 1992, Petersen et al. 2005). Supply chain integration may also lead to internalisation of activities by the focal firm, besides or instead of establishing collaborative arrangements with supply chain firms (Fawcett & Magnan 2002, Stonebraker & Liao 2006). This aspect refers to the issue of vertical integration which is generally absent in the other supply chain concepts mentioned above.

Interest of the building industry in the concept of supply chain integration and related topics

An important stimulus in the building industry's growing interest in the phenomenon of the supply chain has been the *Rethinking Construction* movement, which originated in the UK around the beginning of this century (Egan 1998, Egan 2002a, Egan 2002b). As part of this movement, the search for new and more integrated approaches to the building supply chain has taken on a renewed importance for many firms in the wider building industry (DTI 2003, Holti et al. 2000). Inspired by the British example, the interest in the subject of the supply chain in the *Dutch building industry* followed some years later. In particular, the potential of supply chain integration as a solution to the high levels of waste in building processes, and being able to look collaboratively into cost structures and to long-term mutual financial benefits have been of particular interest (Bouwend Nederland 2009, Building Business 2005b).

Most of the recent efforts to apply supply chain integration and strategic forms of collaboration to the Dutch building industry have taken place in new house building, refurbishment and maintenance, notably by *housing corporations* in co-operation with large builders (Building Business 2007a, Building Business 2007b, Building Business 2008, Building Business 2009, Cobouw 2010, SBR 2009). In addition, a few housing corporations have begun to apply supply chain integration to their renovation programmes as well (Building Business 2010b). The focus has predominantly been on housing corporations as large clients and large builders (Building Business 2010a). However, other types of firms have also expressed their interest and in some cases have started to apply supply chain integration, including SMEs, specialist builders, architects and suppliers (AFN 2009, Doorzicht 2000, SBR 2010a, SBR 2010b).

Parallel to those developments, the industry has shown interest in many *related subjects* such as the rise of the system integrator in building, the idea of viewing building as systems integration, and the application of process integration and advanced business strategies to building (Building Business 2005a, Cobouw 2007, EIB 2005, EIB 2006a, EIB 2006b). As a general solution, many practitioners have pointed towards examples of supply chain integration in *other sectors of industry* as sources of inspiration for the building sector, notably manufacturing, e.g. the automotive and shipbuilding industries (SBR 2007).

1.4 Research objective: specifying a concept of supply chain integration for the building industry

Supply chain integration has been a major influence on the *manufacturing* business for decades. It has, in fact, become a standard around which firms organise integrated delivery processes and organisations (Zailani & Rajagopal 2005). Intense and often global competition, high technological standards and rapidly changing market demands have pressed manufacturers to manage processes throughout the supply chain in an effective and efficient way (Cagliano et al. 2006). The high levels of *alignment and repetition* within the supply chain have led to highly productive and fast operating strategic coalitions of firms (Kim 2006, Kirche et al. 2005, Zailani & Rajagopal 2005). The effect of applications of supply chain integration has been a sweeping alignment of the entire process of design, engineering, logistics, parts manufacture, assembly and finally the delivery of products to the end customer (Hobday et al. 2005). However, in practice supply chain integration has seldom been fully implemented by companies in such a way that the whole supply chain system, from suppliers' suppliers to customers' customers, was integrated. In quite a few cases, companies have found it simply impossible to fully integrate the entire supply chain (Fawcett & Magnan 2002).

In contrast to manufacturing, supply chain integration has not been broadly developed in the *building industry*. Many applications of supply chain management have been aimed at the management of the building materials and long-term arrangements with suppliers. However, a comprehensive approach to the building supply chain, including clients, developers, designers, engineers, contractors, specialists and suppliers has been generally lacking (Briscoe & Dainty 2005). One of the critical phenomena lacking in the building industry is the recognition of a generally accepted *focal company* initiating the integration of the supply chain, as is the case in manufacturing. Often the client organisation has been regarded as the likely proponent for integrating the supply chain (Briscoe et al. 2004, London et al. 1998, London & Kenley 1999). Furthermore, the client may transfer this responsibility to contractors, specialists or suppliers. Alternatively, developers, designers, engineers, contractors, specialists and suppliers may also act like a focal company but usually for a limited part of the whole supply chain (Fernie & Thorpe 2007, Love et al. 2004b, Root et al. 2003)

Parallel to the somewhat unfocused conceptual evolution of supply chain integration in building practice, many *different concepts* have evolved such as partnering (Bygballe et al. 2010), serial contracting (Green & Lenard 1999), multiple project delivery (Miller 1999), strategic procurement (Cox & Townsend 1998), supply chain clusters (Nicolini et al. 2001), supply chain alliances (Dainty et al. 2001b) and supply chain constellations (London 2001). These concepts are indications of the *growing awareness* of the potential of supply chain integration. Potentially, the concepts also refer to different schools of supply chain integration in theory, such as logistics management and purchasing, or in manufacturing practice. However, a unifying concept, much less any school of supply chain integration in the building industry, has been lacking (Briscoe & Dainty 2005).

The path chosen for this thesis is to analytically compare constructs of supply chain integration in general theory to advanced applications of supply chain integration in selected manufacturing industries and to developments in building practice aimed towards supply chain integration (Table 1). The *comparative analysis* is to lead to a theoretical exercise specifying a comprehensive concept of supply chain integration for the building industry.

The *objective of this thesis* is to theoretically specify a comprehensive concept of supply chain integration for the building industry, based on constituting theoretical concepts and advanced applications of supply chain integration in manufacturing, as well as current developments aimed towards supply chain integration in building.

Table 1: Initial direction of the research objective: topics of supply chain integration in theory, applications in manufacturing and direction in building.

Topics of supply chain integration in theory	Examples of advanced applications of supply chain integration in manufacturing	Examples of developments towards supply chain integration in building
Logistics, materials supply	Aerospace e.g. Airbus	Combined delivery of materials to site
Purchasing, transaction costs	Food/grocery e.g. Walmart	Strategic procurement, portfolio management
Supplier management, partnering	Automotive e.g. Toyota	Long-term co-makership with subcontractors
Information sharing, ICT	Computers e.g. Dell	Use of web portals for project management
Production management	Clothing/textile e.g. Zara	Integrated teams on site including specialists
Joint market, co-innovation	Electronics e.g. ASML	Predesigned- and engineered housing concepts by architects and builders
Product development, modular design	Aerospace e.g. Airbus	Prefabricated housing

1.5 Research questions: achieving the objective of a concept of supply chain integration for the building industry

Conceptually, the research objective puts forward the image of the building supply chain envisaged as an *extended enterprise* in which all firms (client, project developer, architect, advisors, contractor, subcontractors, suppliers) virtually operate as 'business units' representing the 'business functions' (marketing, design, engineering, components manufacture, supply, assembly, delivery) of a single 'factory without walls'. In the virtual situation of the extended enterprise, all firms would act as a 'collaborative network of organisational units, regardless of location and regardless who owns them' (Cooper & Rousseau 1999). However, this idea needs to be put into the context of building.

Generally, there are two paths to reducing negative effects of the relatively unstable production situation of building (Ballard & Howell 1998). The first is *transforming the characteristics* of building in order to take advantage of techniques and methods developed in manufacturing, for instance, through simplifying site construction and increasing prefabrication and standardisation. The second is *developing techniques* within building itself that are able to cope with the dynamics of building. The question is what the relationship is between these two paths, leading to the issue of the usefulness of *applying concepts* from other industries, e.g. automotive and aerospace (Gann 1996, Green et al. 2005). When studying the possible transfer and application of 'exotic' concepts to a building context, it is important to learn from other industries how to cope with the characteristics of building without being 'over-simplistic' (Green et al. 2004).

Based on the objective of this thesis, the *central research* question can be stated as follows: How would a concept of supply chain integration for the building industry look like, based on constructs from existing theory, exemplars of advanced applications of supply chain integration in manufacturing, and initial developments towards supply chain integration in the building industry?

This central question has been broken down into six partial questions to be answered by this thesis:

- How is the concept of supply chain integration constituted in current theory? The answer to this question will first be found in theory taken from literature. The concept of supply chain integration has been referred to in generic theory as well as construction management research. The literature covers different theoretical schools of thought and contributing concepts to compose a theoretical framework. The framework will thus need to collect the relevant theoretical insights, existing conceptual thinking, and constituting concepts to be found in literature.
- What supply chain integration practices have been applied in manufacturing? Supply chain integration has been applied in the manufacturing industries for

decades. Comparative studies into a selection of examples in manufacturing will need to reveal what supply chain integration practices can be found, and of what they consist.

- What supply chain integration practices have been applied in building?
  Applications related to supply chain integration can be found in building, too. Case studies into a series of applications must reveal what those applications comprise and what evidence they contain of supply chain integration.
- How do applications of supply chain integration in building versus manufacturing relate?
  Identifying the relationships within and between the applications found in manufacturing and building must increase understanding of the similarities and differences, and identify patterns within the empirical findings. The empirical findings from manufacturing and building will be analysed and compared using within- and cross-case analysis.
- How do the applications of supply chain integrations in practice relate to theory?
  Confrontation of the empirical findings with the existing theory leads to shaping of hypotheses. The hypotheses constitute an emergent concept of supply chain integration.
- What does a concept of supply chain integration in building constitute?
  The hypotheses will point in the direction of the concept of supply chain integration in the building industry, theoretically and practically.

1.6 Limitations and justification: scope and relevance

The thesis focuses on the building industry, including commercial building but primarily *housing*. The focuses is on the *business processes* within the building supply chain, excluding the political and socio-economic context of building. The thesis emphasises the *supply side* of building. This includes the series of firms from suppliers to the client organisation collectively delivering the built object to the actual user. This means the client organisation is regarded as the final stage of the supply chain. Users and other stakeholders in the demand chain have not explicitly been included as distinctive roles. Often building practitioners and researchers regard the user as an integrated part of the supply chain. However, in manufacturing, the end customer is seldom included in the organisation and especially not in the integration of the supply chain. On the contrary, the end customer is the one delivered to and serviced by the supply chain, by means of a sophisticated market approach by the focal firm and the supply chain firms. The thesis therefore focuses on the process of procurement, development, design, production and delivery of built objects and their parts, not including the use and facility management during the life cycle of the built objects.

In the thesis, supply chain integration is regarded as the *integrated management and governance* of the extended delivery process through the supply chain. The management and governance activities observed are connected to the primary activities in the supply chain. However the thesis will not address the integration of primary activities themselves, as explained further in the thesis. Activities are observed as integrated parts of the extended process through the supply chain, rather than distinct activities such as design is often approached distinctively in building theory. This approach is again in line with manufacturing practice, where design is mostly integrated.

Although the theoretical part of the thesis is based on globally available knowledge, the empirical part of the thesis, and most notably the building part of it, applies to the situation in the Netherlands. The thesis is descriptive and explorative by nature. In this respect the thesis is 'broad' rather than 'deep', constructing and giving conceptual meaning to the phenomenon of supply chain integration in building. The thesis will focus on investigating the functioning and the mechanisms of supply chain integration rather than the effects and aims that supply chain firms may pursue applying it, such as cost, time, quality and sustainability effects. The research approach is therefore qualitative rather than quantitative. The thesis is based on a case study approach. However, the applications studied in the selected manufacturing industries are exemplars rather than cases, and used as comparators for the building cases. The manufacturing exemplars as well as the building cases have been selected based on the assumption that the firms involved have actually applied supply chain integration, at least to some extent in the building cases. The thesis is original in the sense that it takes a new approach to supply chain integration from a *multiple theoretical perspective* and a *multiple industrial perspective*. This multiple approach though represents a methodological, analytical and conceptual challenge.

1.7 Structure of the thesis: parts and chapters

This introductory chapter has explained the background, context and objective of this thesis. The research design and approach to achieving the envisaged objective are explained in Chapter 2. This chapter will explain the methodological approach of the research, the research strategy followed, the research methods and techniques applied, and the justification of the research approach and the results.

In the first part of the thesis, the theoretical basis for the research is put forward, starting with *Chapter 3*, which will frame the concept of supply chain integration, based on previous thoughts about the subject and contributing concepts in a building context. In *Chapter 4*, the multiple theoretical framework from four theoretical perspectives is presented: economics, production, organisation and social perspective. Subsequently, theoretical concepts connected to those four perspectives will be presented as a further completion of the multiple theoretical framework. Based on the theoretical framework, *Chapter 5* will present the operationalisation of the

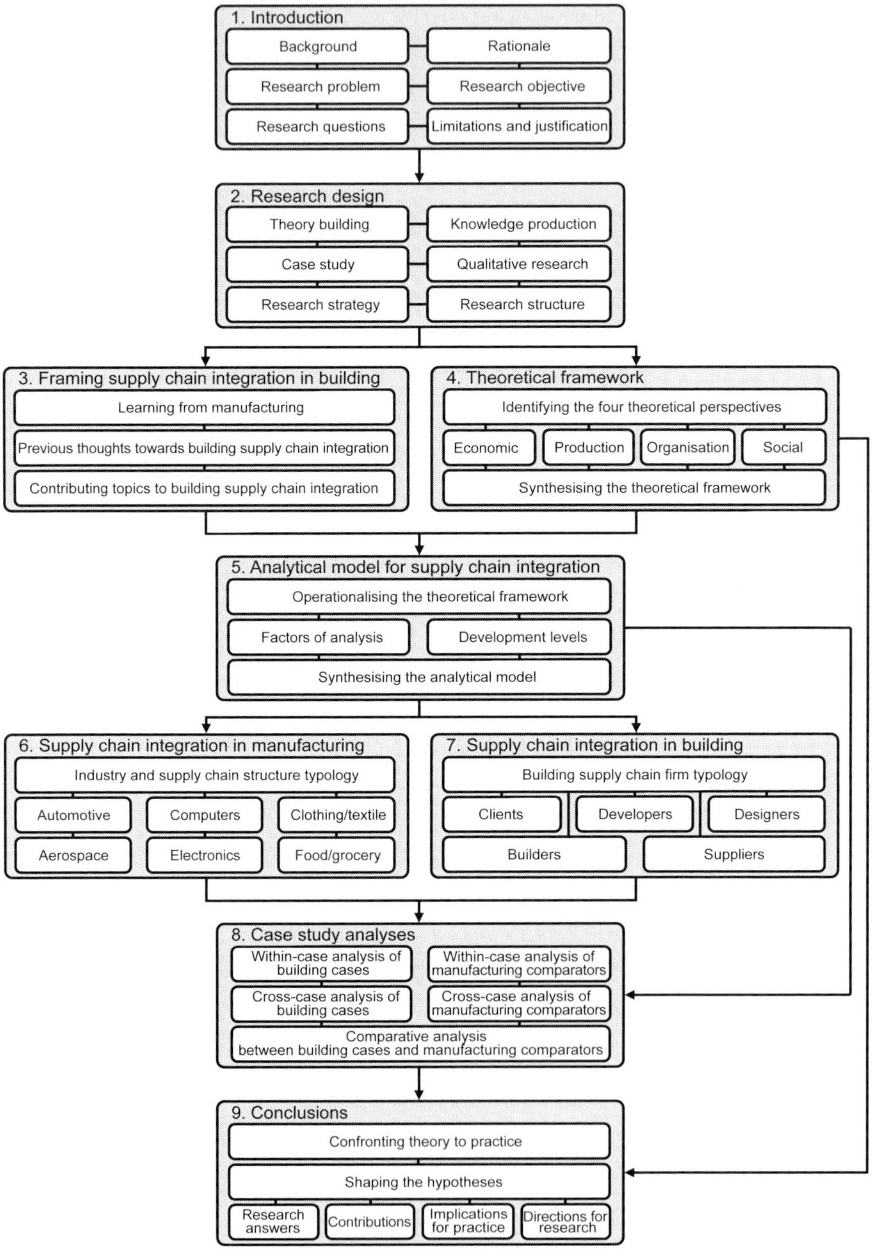

Figure 9: Structure of the thesis

analytical model of supply chain integration that is central to this thesis. The analytical model consists of ten factors of analysis and four development levels of supply chain integration to be used to analyse the empirical findings in the subsequent part of the thesis.

The empirical part of the thesis will start with an exposé of supply chain integration practices in manufacturing, featuring a selection of manufacturing industries as exemplars in *Chapter 6*. After an introduction of industry typologies and characteristics of manufacturing industries, supply chain integration practices from those industries are presented: automotive, aerospace, computers, electronics, clothing/textile and food/grocery. The description of the exemplars of the respective industries will give an explanation of the background, development and state of the art of supply chain integration in manufacturing. The supply chain practices from manufacturing will function as comparators for the building cases. In *Chapter 7*, the building cases of supply chain integration are presented. The cases apply to five firm types along the building supply chain: clients, developers, designers, builders and suppliers. For each type of firm, the current trends and one or more case studies are presented to illustrate the specific application of supply chain integration by the respective firms.

In the synthesising part of the thesis, starting with *Chapter 8*, the empirical findings are interpreted, compared and analysed. First, the findings within and between the building case studies are investigated and compared. Second, the findings within and between the manufacturing industries are investigated, representing a comparator for the building cases. Third, the aggregate findings from the building cases are compared using the manufacturing comparator. Ultimately in *Chapter 9*, the empirical findings are compared to the theory in order to shape hypotheses. Next, the research questions are answered, the contributions of the thesis to theory and method are stated, and the implications for practice are mapped out. Finally, directions for further research are indicated (Figure 9).

1.8 Reflection on key issues and contributions of this chapter to the thesis

This chapter had outlined the thesis. The thesis is aimed at developing a concept of what constitutes supply chain integration in the building industry. The objective is achieved by comparative analysis of multiple constituting theories, examples from selected manufacturing industries, and cases along the building supply chain. The research is a descriptive-explorative hypothesis-generating research. The resulting hypothesis will indicate the contributions the thesis will make to existing theory rather than developing a new theory. The contributions are particularly relevant for the improved fit of existing theory to the particular combination of building industry characteristics, and vice versa.

DANGER
DO NOT ENTER
OVERHEAD HAZARD

2 | Research design

The objective of the thesis is to develop a concept of supply chain integration for the building industry. The development of the concept is approached as a theory-building exercise using case studies, following Eisenhardt's method (Eisenhardt 1989). The resulting concept is to contribute theoretically to the explanation of building and the understanding of what supply chain integration could and should mean to building. The thesis starts with building a theoretical framework using generic theories, and from those theories inducing an analytical model of supply chain integration. This model will be used later in the thesis for the empirical analysis of the selected applications of supply chain integration in manufacturing and building.

2.1 Research approach: contributing to theory and practice

2.1.1 This research viewed as organisational qualitative research

In addition to the objective to contribute theoretically to a concept of supply chain integration for the building industry, the thesis also aims to offer possible solutions and a way forward to improve the building supply chain in practice. Both issues address the organisation of the building supply chain. 'Conducting research in organisational contexts demands that traditional research methods be adapted

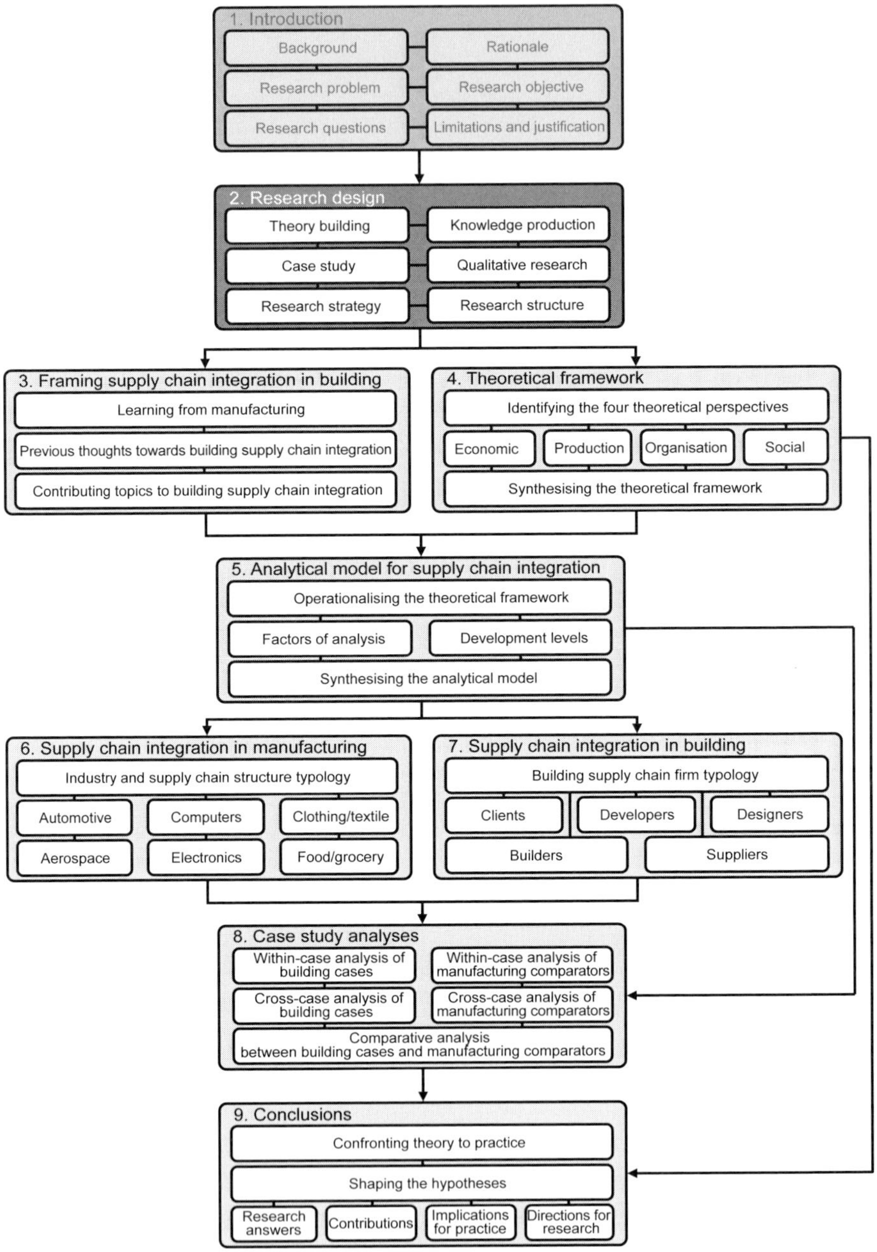
1. Introduction
Background
Rationale
Research problem
Research objective
Research questions
Limitations and justification
2. Research design
Theory building
Knowledge production
Case study
Qualitative research
Research strategy
Research structure
3. Framing supply chain integration in building
Learning from manufacturing
Previous thoughts towards building supply chain integration
Contributing topics to building supply chain integration
4. Theoretical framework
Identifying the four theoretical perspectives
Economic
Production
Organisation
Social
Synthesising the theoretical framework
5. Analytical model for supply chain integration
Operationalising the theoretical framework
Factors of analysis
Development levels
Synthesising the analytical model
6. Supply chain integration in manufacturing
Industry and supply chain structure typology
Automotive
Computers
Clothing/textile
Aerospace
Electronics
Food/grocery
7. Supply chain integration in building
Building supply chain firm typology
Clients
Developers
Designers
Builders
Suppliers
8. Case study analyses
Within-case analysis of building cases
Within-case analysis of manufacturing comparators
Cross-case analysis of building cases
Cross-case analysis of manufacturing comparators
Comparative analysis between building cases and manufacturing comparators
9. Conclusions
Confronting theory to practice
Shaping the hypotheses
Research answers
Contributions
Implications for practice
Directions for research

and adjusted to fit organisational realities' (Swanson & Holton 2005). In order to accomplish this, the research will aim at giving 'answers with meaning' that are holistic, explanatory, context-bound, and qualitative by nature (Leedy & Ormrod 2001). This qualitative approach will lead to research outcomes that tend to be relatively soft, flexible, subjective, regulative and speculative (Silverman 2000).

This thesis aims at giving 'multiple meanings of individual experiences, meanings socially and historically constructed, with the intent of developing a theory' (Creswell 2003). Therefore the thesis is descriptive, explanatory, interpretative as well as exploratory. It is *descriptive* in the way it attempts to reveal the nature of processes, relationships, systems and organisations in the manufacturing and building supply chain. It is *explanatory* of what supply chain integration is in general, and to what extent and in what way it has been applied in manufacturing. It is *interpretative* in the way it provides insight into the nature of supply chain integration, and develops a concept of supply chain integration in the building industry. To conclude, it is *exploratory* in the way it investigates whether the concept of supply chain integration offers an opportunity for the building industry.

The methods used correspond to *qualitative research methods* using informative small samples, observations and interviews (Leedy & Ormrod 2001). The form of reasoning used in the analysis is empirically based and inductive. Communication of the findings will take place in words, narratives, and individual quotations in a literary style, i.e. descriptive analysis (Leedy & Ormrod 2001). The merit of the qualitative research approach in this thesis is that it provides a means to judge and evaluate the effectiveness and causality of particular policies or practices (Leedy & Ormrod 2001). In particular, it allows testing of the validity of assumptions, claims, theories, or generalisations within the *real world context* of building (Robson 1993).

2.1.2 This research viewed as knowledge production

Views on management research have given rise to the concept of knowledge production, in particular the development of problem-solving knowledge which is applicable to practice, such as *Mode 2 knowledge production* (Gibbons et al. 1994). In contrast to Mode 1 knowledge, which is abstract and universally valid, Mode 2 knowledge is applied, contextually embedded, 'more socially accountable and reflexive' (Gibbons et al. 1994). Mode 2 knowledge production aims at designing 'solution-oriented research products' rather than merely deducing 'analysis-based explanations' (Van Aken 2005). The characteristics of Mode 2 knowledge production contribute to the aim of this thesis to contribute to theory as well as to practice. In this respect, Mode 2 knowledge production is related to 'real world enquiry', including 'problem solving rather than just gaining knowledge' (Robson 1993).

This thesis can be qualified as Mode 2 knowledge production. In particular, the following attributes of Mode 2 knowledge are useful contributors to the aim of the thesis: 1) *knowledge production in the context of application* (problem solving organised around a particular application), 2) *transdisciplinarity* (consensus and integration of different skills in a framework to guide problem solving), 3) *heterogeneity and organisational diversity* (composition of the problem solving framework over time), 4) *social accountability and reflexivity* (contextual awareness and consideration) (Gibbons et al. 1994). These attributes support the potential of this thesis to lead to a theoretical contribution as well as practical insights aiming towards the integration of the building supply chain (Handfield & Nichols 2002).

2.1.3 This research viewed as theory building

The views outlined above will lead to the development of a concept of supply chain integration in the building industry. This implies a *contribution to existing theory*. The elements of this thesis leading to the theoretical contribution include constructs from existing theory related to supply chain integration, existing thoughts and concepts in building theory and practice related to supply chain integration, and empirical findings on applications of supply chain integration in manufacturing as well as in building. This approach is aimed at 'constructing reality and thus raises methodological challenges caused by contextualised observation and explanation of perceptions and behaviours of firms and individuals with whom they interact' (Patton 2002). This is largely influenced by the viewpoint taken by the researcher and the appearance of the research subject in theory and practice. Therefore, epistemologically, the thesis takes a *constructionist* viewpoint. Second, the research is very much an observation (i.e. exploration and interpretation) of what is found in literature and particularly in practice, and based on that, gives factual meaning to the findings. Therefore, ontologically, this thesis takes an *empirical* position.

This thesis adopts an existing body of theory and aims to contribute to it rather than develop new theory. A major problem in *theory building* is that most empirical generalisations are logically consistent with a good many theories, and different theories lead to many of the same predictions. Therefore, the theory building within this thesis is expected to represent *extensions to existing theory*, particularly aimed at conceptualisation of the phenomenon of supply chain integration in the building industry. The process of this theory building will be dominated by description and exploration, and thus it cumulates to theory along a sequence of repeated extension, validation and refinement. In accordance with Weick's description (Weick 1989), the theory building can be best viewed as a 'process of imagination disciplined by evolutionary processes analogous to artificial selection; a process of thinking and trial'. In the end, it is important that it is a process that is 'well-defined, well-argued and well-executed' (Snijders & Vos 2007).

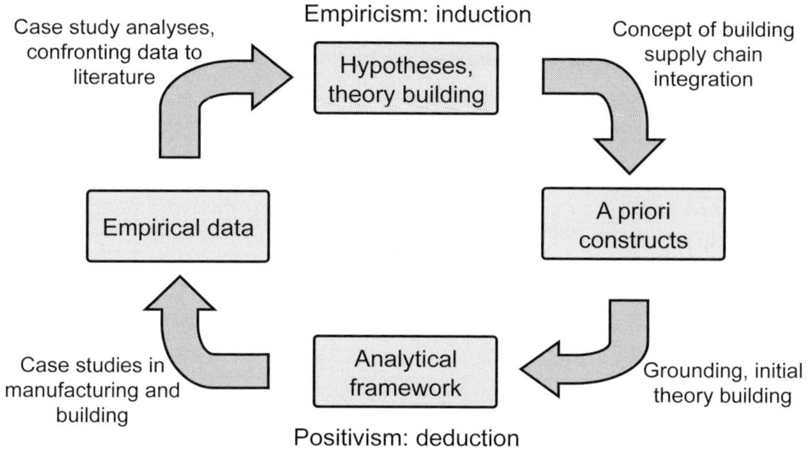

Figure 10: Deduction and induction as applied in this research. Adapted from Love et al. (2002a)

2.2 Research strategy: theory building from case study research

This thesis will follow a combined *deductive* and *inductive* strategy towards the envisaged theory building (Wacker 1998) (Figure 10). First, a collection of theoretical assumptions from general theory will exist as a priori theoretical constructs leading to the deduction of a specific analytical framework for supply chain integration. The a priori theoretical constructs provide initial grounding for the theory building further during the research process. Next, the thesis will follow a case study approach using the case study approaches of Yin (1989), and particularly Eisenhardt (1989). Yin's multiple case study approach will mainly be used to induce empirical findings from the different sets of *case studies*. Eisenhardt's theory building approach will play a dominant role to generate hypotheses from the empirical findings.

The case studies consist of multiple sets of studies of various applications of supply chain integration in manufacturing and in building. The *manufacturing cases* are first used to relate the theoretical constructs and particularly the analytical framework to the practical context of manufacturing. Next, the manufacturing cases function as intermediates between the theory and the building cases, and as such they will not function as case studies but as *comparator studies* for the *building cases*. While the manufacturing comparator studies in essence represent secondary descriptions of various segments of manufacturing, the *building cases studies* are to find direct empirical evidence of supply chain integration in a building context. The buildings cases are analysed and compared to the manufacturing comparators and to the *theory* later on in the thesis. This composite character of the thesis and the varied constellation of the cases require the data collection and analysis *methods* to be flexible, enabling access at various stages of the supply chain and allowing a range of data gathering and assessment techniques (Seuring 2008). Therefore, in this thesis a multiple case study approach is used in order to increase the number of instances data is gathered in different settings, and hence permit drawing wider conclusions and shaping useful hypotheses covering the multi-staged and multi-faceted phenomenon of supply chain integration (Figure 11).

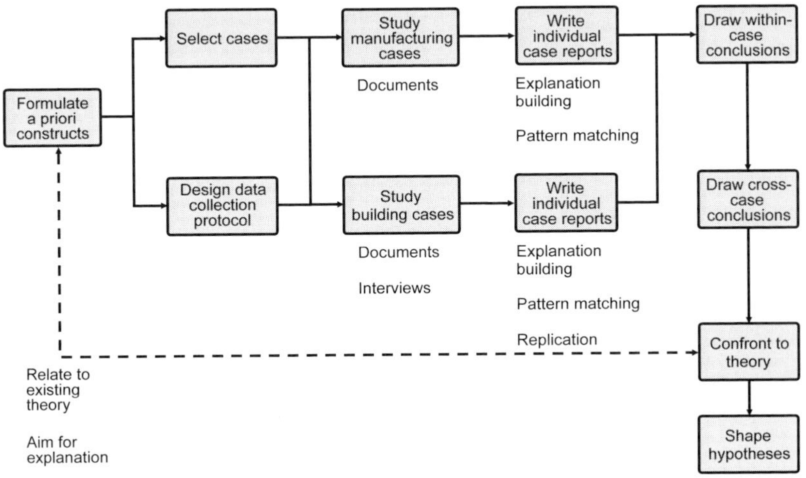

Figure 11: Multiple case study approach to theory building as applied in this thesis. Adapted from Yin (1989) and Eisenhardt (1989)

This thesis follows the above directions for theory building as an 'evolutionary process' starting and ending with theory, and thereby contributing to theory. The process leading to this theoretical contribution follows Eisenhardt's hypotheses generating process (Eisenhardt 1989) (Table 2). This process is further extended in the structure of the research given below, anticipating the following chapters.

Table 2: Process of theory building from case study research in this thesis (Adapted from Eisenhardt 1989).

Step	Activities	Reason
Getting started	• Definition of research objective and questions • Exploring a priori constructs of supply chain integration in theory • Selecting theoretical perspectives on the matter, without postulating hypotheses	• Focusing the research • Providing grounding and shaping the initial design of the theory building • Retaining theoretical flexibility
Selecting cases	• Specifying sectors and firms in manufacturing and building for the cases • Analytically informed, not random sampling	• Reducing extraneous variation and supporting generalisation • Selecting cases likely to sustain and extend the emergent theory
Crafting instruments and protocols	• Multiple data collection methods to collect primary and secondary data from the cases • Multiple sources and perspectives on multiple types of data	• Achieve triangulation of the evidence to strengthen grounding • Divergent and synergistic views on case evidence to strengthen grounding
Entering the field	• Parallel data collection and analysis in manufacturing and building cases • Flexible and semi-structured data collection	• Speed of the analysis and possible adjustments to the data collection method • Being able to benefit from shifting data and emerging themes during data collection
Analysing data	• Within-case analysis, understanding, explanation building • Cross-case analysis, pattern matching, applying multiple techniques	• Taking advantage of rich data for preliminary theory generation • Looking beyond initial impressions, gaining more understanding, better grounding
Shaping hypotheses	• Iterative tabulation of evidence from the cases • Identifying replication logic across cases • Explaining and sustaining causality of evidence	• Supports construct validity • Confirms and extends the emerging theory • Supports internal validity
Enfolding literature	• Confrontation of empirical results with theory, literature	• Improves further validity, definition, generalisation, and theoretical level
Reaching closure	• Theoretical saturation, end of iterating theory and data	• Ends the process when incremental improvement of the theory becomes minimal

2.3 Structure of the research: research stages and activities

2.3.1 Part I: Theoretical basis

Exploring the field (Chapter 3)

First, the concept of supply chain integration in the building industry is framed. This includes an exposé of *previous indications* as to the lessons that building could supposedly draw from manufacturing, and *previous thoughts* about supply chain integration in building. In addition, a *review* is presented of various contributing concepts that have emerged in recent decades in an attempt to integrate the building supply chain, such as strategic procurement and partnering. Following Eisenhardt's method, the collection of those lessons, thoughts and concepts represents a first set of a priori *empirical constructs* (Eisenhardt 1989). The exploration and review of the constructs indicate *gaps* in the current understanding of supply chain integration in the building industry, and provide initial *grounding* for the theory building of the research while retaining theoretical flexibility.

Literature study (Chapter 4)

The literature study of this thesis will investigate and provide a theoretical view on the concept of supply chain integration from *four theoretical perspectives*: social theory, economic theory, organisational theory and production theory. From those four domains, a priori *theoretical constructs* will be collected and reviewed. The reviewed constructs provide theoretical 'building blocks' for the theory building further in the thesis, and the formation of the *analytical framework* in Chapter 5. Further to the empirical constructs identified in Chapter 3, the theoretical constructs give direction for the contribution to existing theory at which this thesis aims, i.e. the development of a concept of supply chain integration for the building industry.

Analytical model (Chapter 5)

The theoretical framework of Chapter 4 will lead to the formation of the analytical model for supply chain integration, which will function as the *initial representation* of the concept of supply chain integration to which the thesis will lead. This model will also function as the *framework for analysis* of the case studies in Chapters 6 and 7. The model will consist of *factors* of analysis and development *levels* of supply chain integration derived from the previously introduced empirical and theoretical constructs. As a result, a two-dimensional model will arise of analytical factors and development levels, which will further be used for the description and the analysis of the cases in the following chapters.

2.3.2 Part II: Empirical analysis

Unit of analysis

The unit of analysis of the empirical research is the *focal firm* in the supply chain applying supply chain integration strategies towards neighbouring activities or firms in the supply chain. In the *manufacturing* cases, the focal firm will generally be the manufacturer or central firm leading the manufacture, assembly and delivery of the final product to the end customer. In the *building* cases, the focal firm may refer to different types of firms in the supply chain, depending on the firm taking the initiative towards supply chain integration, e.g. the client organisation or the primary contractor. The building cases will be categorised based on the types of focal firms selected for the case studies, including the part of the supply chain that particular firm has integrated, and the strategic arrangements established with other firms in the supply chain.

Case selection

The selection of the cases to be studied has been *analytically informed* to reduce variation as much as possible following Eisenhardt (1989). In addition, the cases selected were likely to sustain and extend the theory building of the research. The *manufacturing cases* were selected using an industry and supply chain typology. The manufacturing cases represent three basic supply chain structures and as a result, give a relatively broad view on what supply chain integration in manufacturing implies. Few of the manufacturing cases have been previously identified in research and practice as potential examples for building. The *building cases* were selected from the Dutch building industry based on indications from publications and media that these firms are of particular interest in the way they have made an attempt to apply supply chain integration strategies. In addition, the building cases were selected to cover collectively the building supply chain from client organisation to supplier. The selection of both manufacturing and building cases is further explained in Chapters 6 and 7 respectively.

Comparator studies in manufacturing (Chapter 6)

The case studies in manufacturing are based on *secondary data*, e.g. writings about supply chain integration strategies and techniques as applied by manufacturers in a number of manufacturing industries. These case studies do not represent the actual case studies of the thesis. Instead, they function as *intermediates* and *comparators* between the theory and the actual case studies, i.e. the building cases. The manufacturing industries studied include the automotive, aerospace, computers, electronics, clothing/textile and food/grocery industries. The firms studied in these industries respectively represent leading examples of supply chain integration. Accordingly, the firms studied function as examples within the industry sectors studied, rather than as actual case studies.

Case studies in building (Chapter 7)

The case studies in building are studied based on *primary and secondary data* of a selection of firms in the Dutch building industry, and their application of supply chain integration strategies. The firms represent *five firm types* along the building supply chain, from client organisations, to project developers, designers, builders and suppliers. In all cases, the firms have applied strategies or developed specific concepts that relate to a form or a part of supply chain integration. The procedures and techniques for the case studies have been described in a case study protocol (see Appendices).

Data sources

The data sources include primary data sources as well as secondary data sources. Primary data sources refer to the focal firms investigated in the building cases, including *interviews* and *internal documentation*. Secondary data refers to *external documentation*, such as news articles and internet resources about the application of supply chain integration by the focal firms studied, both in the manufacturing cases and in the building cases[1].

Data collection

The three most important kinds of qualitative data or sources of evidence are interviews, observations and documentation (Patton 2002, Yin 1989). In this thesis, interviews and documentation in particular have been used. The use of *multiple data sources* has led to increased reliability (Yin 1989). The use of *multiple data collection methods* has led to stronger substantiation of the constructs underlying this thesis as presented in the preceding chapters (Eisenhardt 1989).

Case study database

The data collected in for this thesis was stored in a case study database for further analysis. The data was archived in a structured way based on the specific industry sector, the type and details of each firm. The *data structuring* included the classification of the particular concepts and strategies applied by the focal firm in each case, and the collaborative arrangements established with supply chain firms (see Chapters 6 and 7 and Appendices).

2.3.3 Part III: Synthesis

Data analysis

The data analysis took place according to the following steps (Leedy & Ormrod 2001): organisation of details per case, interpretation of single instances (e.g. comparing documents to interviews), identification of patterns, and synthesis and generalisation (i.e. synthesis into an overall portrait of each case and across cases). The *coding* used for the analysis of the building cases was derived from a process of open and axial coding of the manufacturing comparator studies. The open coding process consisted of identifying first order themes from the studies. The following axial coding process consisted of grouping related, first-order themes into larger, second-order themes (Miles & Huberman 1994, Strauss & Corbin 1998) (see Appendices). Next, *data reduction* in the building case studies was accomplished by filtering and coding quotations from the rough data. This was done manually

1 The manufacturing cases must be observed as exemplars rather than genuine case studies, functioning as comparators for the building cases that represent the actual case studies of this thesis.

and with additional computer assistance using qualitative data analysis software. This combined approach was followed in order to secure sufficient 'feeling' with the findings and the quality of analysis of those findings. For further analysis, additional techniques were used, including narrative analysis, pattern matching and explanation building (Yin 1989).

Table 3: Overview of the thesis and research structure.

Part	Step	Research activity	Outcome	Aim	Chapter
Part I: Theoretical basis	Exploring the field	Desk research	Framing the concept of building supply chain	Explore	3
	Literature study	Desk research	Theoretical buildings blocks	Describe	4
	Analytical model	Desk research	Theoretical model, analysis framework	Integrate	5
Part II: Empirical analysis	Comparator studies in manufacturing	Desk research	Studies using secondary data	Describe, specify	6
	Case studies in building	Case study research	Studies using primary and secondary data	Describe, explore	7
Part III: Synthesis	Case study analysis and interpretation	Within-case and cross-case analysis	Review of results, relating variables, pattern matching	Reduce, explain, interpret	8
	Conclusions	Extrapolate results, shape hypotheses	Presentation of theory, model and implications	Integrate, evaluate	9

Case study analysis and interpretation (Chapter 8)

Based on the data analysis, further explanations were searched for by means of within-case and cross-case analyses. The *within-case analyses* refer to individual analyses of each of the case studies in building and the comparator studies in manufacturing. The analyses of both types of studies have been done in a similar way to increase comparability. The *cross-case analyses* refer to comparisons between the individual studies in manufacturing and in building, but more particularly between the manufacturing comparator studies and the building case studies. Following Eisenhardt's method, this has been done by iterative tabulation, comparison and explanation building to improve grounding for the theory building (Eisenhardt 1989).

Conclusions (Chapter 9)

To conclude, the empirical findings were confronted with the existing theory and the constructs postulated in the first part of this thesis. Via a process of further tabulation and iterative permutation, a comprehensive set of hypotheses had been formulated, i.e. extensions of existing theory. The extensions represent the theoretical contribution of this thesis and thus mark the closure of the theory building process of the research according to Eisenhardt (1989). Finally, the research questions are answered, the contribution of the thesis to theory and method are summed up, implications for building practice are specified, and directions for further research are given (Table 3).

2.4 Justification of the research approach

2.4.1 Validity

Construct validity

Validity refers to 'truth interpreted as the extent to which an account accurately represents the phenomena to which it refers' (Silverman 2000). Construct validity is the measure of *truthful representation* of the concepts under investigation (Emory & Cooper 1991). In the case of qualitative research such as this, validity has to do with the *information richness* of the cases selected rather than with sample size (Patton 2002). In this thesis, supply chain integration as the concept under investigation is first underpinned by a vast body of theory, resulting in an analytical model. Next, the concept is explored by multiple studies within the contexts of manufacturing and building. The aim is increased richness of information and hence improved construct validity.

Internal validity

Internal validity refers to causal relationships between variables in the research (Miles & Huberman 1994). Internal validity allows accurate conclusions to be drawn about the *causality* within the data. To assure internal validity, it is important to consider *alternative explanations* and apply *triangulation*. This was done by using multiple research methods, multiple theories, different data sources and different analysis techniques, under the assumption that they all converge to support *coherent findings* and thus lead to internal validity (Brewer & Hunter 1989, Patton 2002).

External validity

External validity is the extent to which research findings apply to situations *beyond the research itself*. The main concern related to case study research and therefore this research as well is that the external validity of the results is often questionable (Gummeson 1991). Particularly if only a few cases are studied, it can be difficult to declare the results *applicable to broader situations* or even to similar cases. This research aims to achieve external validity by extensive description of multiple case studies (Eisenhardt 1989). The resulting empirically driven discussion of the research findings was used to relate the findings to a *wider body of thoughts* and thus construct a 'clearer reality' (Stake 1995). In the conclusion of the thesis, a central role is reserved for the process of relating the research outcomes to existing theory in order to increase the external validity.

2.4.2 Reliability

Reliability refers to the degree of *consistency* with which conclusions are attributed to the same instances by different observers, or by the same observer on different occasions (Silverman 2000). The issue of reliability is closely linked to the issue of *replication*, which is defined as the scope to which the research findings can be replicated beyond the cases under study (Emory & Cooper 1991). Reliability therefore refers to the ability of other researchers to carry out the same study and achieve similar results. The measures used to increase the reliability in this research are the case study protocol and the coding used to assure that the case studies and their results are verifiable (see Appendices). The protocol is a particularly essential element to ensure the reliability of case study research (Yin 1989). This protocol contains the field and analysis procedures so that the results can be replicated.

2.4.3 Generalisability

In theory building research, the issue of generalisability has often been a major concern, particularly in the case of qualitative research. Following the *replication logic* of Yin (1989), subsequent case studies enable the researcher to test, affirm or falsify, and thereby develop a valid theory. The theory that is ultimately formulated must then become the vehicle for generalisation based on the comparison to external views and findings. Therefore, this thesis has not endeavoured to build a widely acceptable new theory, but rather to contribute and rely on existing theory for the phenomenon under investigation, i.e. supply chain integration in the building industry. This was achieved via a process of *analytical generalisation* (Smaling 2003). This type of generalisation makes no claim to statistical representativeness, but instead assumes that the results contribute to an existing theoretical explanation of the phenomenon (Flyvbjerg 2006). In this way it particularly helps qualitative research not to limit its value. Central to this approach is the *force of example* on which this thesis has also relied. The subsequent *mental extrapolation* of the research results has allowed an understanding of the diversity and heterogeneity of the matter at hand beyond the instances observed (Maxwell 2007). Following Eisenhardt (1989), tying the *emergent theory* to existing and generic theory has enhanced the generalisability of the outcomes of the theory building process of this thesis (Amaratunga & Baldry 2001).

2.5 Reflection on key issues and contributions of this chapter to the thesis

This chapter has explained and underpinned the research as qualitative research, characterised as knowledge production leading to theory building. The objective of the thesis is not to build a new and generally valid theory but rather to contribute to existing theory or extend existing theory specifically for the conceptualisation and application of supply chain integration in the building industry. This is attempted by means of case study research, inductively building the theoretical contribution of the thesis in a descriptive and explorative manner.

PART I:

THEORETICAL BASIS

FlowTrack

3 Framing the concept of supply chain integration in building

This chapter will frame the concept of building supply chain integration, based on previous thoughts and concepts that have arisen in building practice and theory. First the idea of learning from other industries, notably manufacturing, will be addressed. The differences between building and other industries will be acknowledged and addressed, which may hinder the transfer of concepts from those sectors to building. Second, previous thoughts pointing towards the merits as well as the doubts about supply chain integration in building will be reviewed. Third, a number of topics existing in building practice and literature, and their potential contribution to the concept of supply chain integration will be presented. The collection of the conceptual contributions in this chapter provides initial grounding to the concept of supply chain integration for the building industry, to be further developed theoretically in the following chapters.

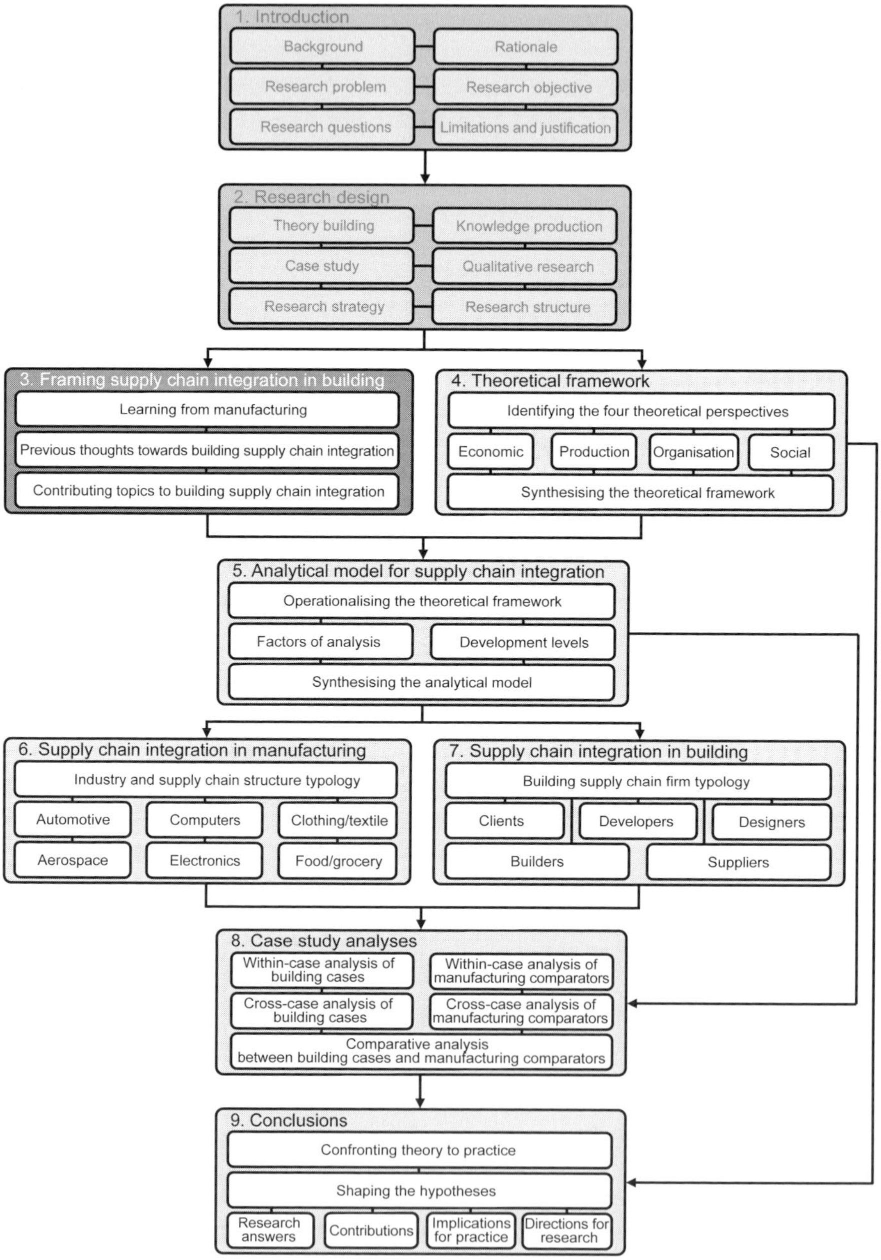
1. Introduction
Background
Rationale
Research problem
Research objective
Research questions
Limitations and justification
2. Research design
Theory building
Knowledge production
Case study
Qualitative research
Research strategy
Research structure
3. Framing supply chain integration in building
Learning from manufacturing
Previous thoughts towards building supply chain integration
Contributing topics to building supply chain integration
4. Theoretical framework
Identifying the four theoretical perspectives
Economic
Production
Organisation
Social
Synthesising the theoretical framework
5. Analytical model for supply chain integration
Operationalising the theoretical framework
Factors of analysis
Development levels
Synthesising the analytical model
6. Supply chain integration in manufacturing
Industry and supply chain structure typology
Automotive
Computers
Clothing/textile
Aerospace
Electronics
Food/grocery
7. Supply chain integration in building
Building supply chain firm typology
Clients
Developers
Designers
Builders
Suppliers
8. Case study analyses
Within-case analysis of building cases
Within-case analysis of manufacturing comparators
Cross-case analysis of building cases
Cross-case analysis of manufacturing comparators
Comparative analysis between building cases and manufacturing comparators
9. Conclusions
Confronting theory to practice
Shaping the hypotheses
Research answers
Contributions
Implications for practice
Directions for research

3.1 The idea to learn and transfer supply chain concepts from other industries to the building industry

3.1.1 General lessons from manufacturing as potential solutions for building

When discussing the building supply chain, reference is often made to other sectors of industry, notably manufacturing. However, the unique combination of characteristics in the building industry has provoked a debate about the usefulness of comparing the building industry to those other industries where the levels of *repetitiveness* and *integration* of the delivery system through the supply chain are normally higher. The concept of repetitiveness and specifically, the development of concepts and methods that increase the level of repetitiveness, are often suggested to promote the improvement of building practice. The usefulness of adopting concepts from manufacturing to improve building practice has been discussed previously (Gann 1996, Voordijk & Vrijhoef 2003). Increasing the repetitiveness of building has been achieved for instance by multi-project and portfolio approaches and developing and delivering integrated and standardised building solutions (Blismas et al. 2004b, Martinsuo & Lehtonen 2007). Implicitly, these attempts seem to have endeavoured to redefine traditional building towards a manufacturing format. In short, an important lesson that building can learn from manufacturing is its effective and efficient forms of integrated and repetitive organisation and management of processes leading to unique products under time constraints (Winch 2006).

The case of learning from supply chain management in aerospace

The practice of supply chain management in building seems less far advanced than in the aerospace industry. Although, the emergence of concepts like prime contracting and framework agreements in the building industry has provided a more supportive climate for supply chain management than has traditionally been the case. However, the progress of these developments depend upon *continuity of workload* under such arrangements, i.e. repetitiveness (Green et al. 2005). Providing that a regular workflow can be achieved, those arrangements may well enable *long-term collaborative relationships* with key suppliers, comparable to the established practice within the aerospace industry. Such trends are already observable in building, although the influence of supply chain management on the building industry at large may remain small (Green et al. 2002). An explanation could be found in the *differences* in terms of institutional context, structure and technological intensity. In contrast to the building industry, the aerospace industry operates globally and it has experienced extensive consolidation dominated by a small number of global players. Aerospace companies operate within such wide networks of *global interdependency* that *collaborative working* has become a 'commercial imperative' (Green et al. 2004). The imperatives of international competition are among the factors that have obliged the industry to operate in a highly collaborative manner. 'It is the shared ethos of mutual dependency that provides the platform for effective supply chain management' (Green et al. 2002).

3.1.2 Differences between industries as a potential hindrance transferring concepts

Despite the potential benefits of adopting manufacturing concepts that could improve the building supply chain, differences between the building industry and the originating industries could hinder that transfer. Translation or transformation of these 'exotic' concepts is generally needed when transferring and applying them to a building context to avoid being 'over-simplistic' (Green et al. 2004). This calls for a good understanding of the differences in industry *structure, culture and management traditions*. In the building industry, the management challenge is mainly the industry's project focus, together with the relatively informal culture, and the fragmented structure of the industry as a whole and the production situation in particular (Table 4).

The characteristics of industries vary from one to the next. The production system of each industry has been shaped by that industry's characteristics and history. Project production systems in *project-based industries* such as building are aimed at a product mix that is 'one of a kind or few', process patterns are 'very jumbled', process segments are 'loosely linked', and management challenges are dominated by 'bidding, delivery, product design flexibility, scheduling, materials handling and shifting bottlenecks' (Schmenner 1993). The *fragmentation* of the building industry has been identified for decades as a major focus of the complaints about the state of practice (Turin 2003), reflected most characteristically by the predominantly *one-off approach* in building projects, or 'unique-product' production (Drucker 1963). On the other hand, the characteristics of building resemble other project-based industries, but it is its *specific combination of characteristics* that makes the building industry unique, and which presents a challenge to applying such concepts as supply chain integration to the building supply chain (Segerstedt & Olofsson 2010).

Table 4: Characteristic differences between building and other industries (Vrijhoef & Koskela 2005c)

	Building	Other technology-driven sectors	Non-technology sectors
Culture	Informal	Formal	Dynamic
Structure	Fragmented	Consolidated	Integrated
Management	Project driven	Process driven	Customer driven

The *culture* in building is relatively polymorphous and heterogeneous, a result of the relatively large number of different firm types and sizes and the organisational configurations of projects. The culture within building is a typical 'project culture' and is relatively informal compared to the frequently more formal 'corporate culture' dominant in other industries such as manufacturing. The high status of projects explains the existence of two cultural identities within the building industry: the corporate culture of firms, and a distinctive culture within each separate project. The fragmented production system, the strong influence of project culture, a relatively weak corporate culture, and a lack of shared values in the supply chain have been regarded as reasons for the relatively limited possibilities to achieve value for the client effectively and efficiently (Riley & Clare-Brown 2001). Project characteristics differ noticeably across project-based industries. Besides the *product aims* of achieving satisfactory quality and customer value, usually the normative *project aims* in building projects are the budget and the completion date. The project success measure is cost, and completing the project by the scheduled date is generally the most important scheduling objective (Tukel & Rom 1998).

The *project management* function logically plays a central role in building. The dilemma is that the larger and more complex the project is, the more empowered the project manager must be to exercise control and authority, but at the same time, the more he or she must delegate and trust the project team (Hammuda & Dulaimi 1997). The standards and models used in the building industry are relatively basic in comparison to many other industries (Cooke-Davies & Arzymanow 2003). In terms of *quality management*, differences have been suggested between industries related to the level of quality management implementation and quality output performances. In building, the attitude tends to be oriented towards conformance to contractual specifications rather than gaining additional financial benefits or competitive strength from quality improvement. As a result, building has been deemed to be less commercial and market-responsive, but rather oriented more towards production and getting the work done on time and within budget (Lai & Cheng 2003).

3.2 Previous thoughts pointing towards supply chain integration in building

3.2.1 Quasi-firm and extended enterprise

Previous thoughts have emerged within building practice and literature, pointing in the direction of supply chain integration. The 'quasi-firm' was a seminal idea first introduced by Eccles (1981), promoting *strong linkages* between firms involved in a building project, particularly main contractors and subcontractors. The quasi-firm points towards the notion of those firms behaving as 'one firm'. This raises the issue of core competences of firms together making up an 'extended enterprise' in a resource-based view (Barney 1991, Prahalad & Hamel 1990). The extended enterprise in building supply chains implies a higher level of integration between firms. In order to achieve higher levels of supply chain integration, there is a need to strengthen *inter-firm relationships*, achieve mutual benefits and build trust in the supply chain (Dainty et al. 2001a). In order to accomplish this, firms involved need to take away the often mentioned ingrained barriers of traditional relationships and the adversarial culture in building practice. An important issue of the extended enterprise in the building supply chain is the effective *division of operations* and the efficient *allocation of specialised tasks* among the relatively many specialist firms involved in building projects, and next the *coherent coordination* of these operations, tasks and firms. This is particularly an issue in the building industry because of the relatively high share of *SMEs* (Briscoe et al. 2001, Dainty et al. 2001b, Millett et al. 2000). This calls for intensive *cooperation* and *communication* in the network of firms, supported by actual or either virtual inter-firm coordination measures (Kornelius & Wamelink 1998). Therefore, it is important to distinguish tight or actual *couplings* between firms in individual projects from loose or virtual couplings in the permanent network of firms within the industry, viewed as a 'loosely coupled system'. In the case of firms in the industry network joining together in building projects, the permanent loose couplings will get tighter temporarily for the duration of the project (Dubois & Gadde 2002). The view on the supply chain as a single firm is useful for the concept of supply chain integration, however repetitiveness is only implicitly promoted as an additional topic. However, the idea of the single firm has mainly applied to contractors, subcontractors and suppliers, generally excluding additional firm types such as project developers and designers. The idea is tends to focus on inter-firm relations on an economic level within the industry, underexposing production and organisation aspects on the management and governance level of the supply chain.

3.2.2 Supply chain management

Supply chain management has been another important field of research in construction management literature and practice, and often coined as a potential source of improved performance for the building industry (London & Kenley 2001, O'Brien et al. 2002, O'Brien & Fischer 1993). Supply chain management has been mentioned mainly as an opportunity to reduce costs and time by improving *logistics*. It has also been seen as a means for more effective and efficient *product development* and *marketing* of building materials (Olsson 2000). Another direction for supply chain management has been towards establishment of *partnering* arrangements within the supply chain (Fernie & Thorpe 2007). The application of supply chain management to building has generally been observed to be complex because of the multiple implications for building practice in relation to the lack of *central control* in the supply chain. However, centrally controlled supply chains are not merely directed towards obvious aims such as *minimising transaction costs*, but also towards enhancing the *transfer of expertise* and systematic *feedback* on planning, design, building and maintenance between parties, and striving for joint *value maximisation* (Voordijk et al. 2000). These ideas about supply chain management connect well to supply chain integration, and have shown applicability to building practice. Those applications have generally regarded the main contractor as the focal firm in the supply chain, which though is not necessarily always the case in building practice.

3.2.3 System integrator

The increase of central control in the supply chain has considerable consequences for the *power balance* among firms in the supply chain (Ireland 2004). As the level of control increases, the level of *autonomy* in the supply chain will decrease. This gives rise to the idea of the system integrator taking control of the integrated supply chain. For instance, client organisations who would take on the role of system integrator could *involve teams of building partners* for longer periods of time, e.g. through prime contracting and framework agreements. Contractors taking on this role could *mobilise subcontractors* and integrate all activities needed to deliver complete products to the market. Alternatively other firm types with the power and capabilities could engage in *strategic collaboration* with supply chain partners in order to deliver integrated products to the market collectively. In the various sectors within building, the possibilities and implications of exercising the role of system integrator through centrally controlled arrangements are different from a demand perspective and from a supply perspective (Table 5).

Table 5: Differentiating between demand and supply perspectives on the role of the system integrator (Vrijhoef & Koskela 2005a)

	Demand perspective	Supply perspective
Housing	Individual clients generally lack power to integrate the supply chain; large housing corporations may develop a supply chain strategy	Contractors or suppliers may engage in strategic collaboration to develop and introduce pre-engineered housing concepts to the marketplace
Commercial building	For instance, real estate developers or large companies may involve integrated supply chains for their real estate stock	Contractors or suppliers may join to develop multiple commercial buildings; however, commercial risks and variability of design, etc. may be too high
Civil	Public or semi-public clients in particular may introduce portfolio procurement strategies for multiple projects	Contractors or suppliers may deliver multiple integrated projects, but mostly initiated by the client, not through pre-engineered concepts

Demand system integrator

Traditionally, client organisations have played a *dominant role* in initiating building projects (Cherns & Bryant 1984), in making the initial decision to procure building works and in the way in which procurement takes place (Briscoe et al. 2004). A lack of *continuity* of relationships may hinder reaping the full advantages of long-term collaboration and the transfer of experience and knowledge across projects (Bresnen & Marshall 2000a). In order to be a demand system integrator, the client represents the entire group of users, financiers and other stakeholders involved. Clients who have the power to shift their *procurement* strategies are in a position to align the supply chain effectively and implement their role of system integrator successfully (Cox & Ireland 2001). Innovative procurement strategies may include performance and financial incentives, sharing 'pain and gain', and relying on the positive effect of long-standing relationships (Khalfan et al. 2005, Pryke 2004). Those incentives tend to *reinforce relationships and commitment*, and foster trust for longer periods of time, which apparently cannot be achieved through a one-off approach to single projects (Bresnen & Marshall 2000b). Some advanced clients have been able to create multi-project environments through a *portfolio approach* aimed at increasing repetition and creating similarities between multiple projects, and thereby increasing the degree of certainty and stability (Abdullah & Vicridge 1999, Blismas et al. 2004a, Cox & Townsend 1998).

Supply system integrator

Although the client and the demand are critical and initiating factors in building, parties on the supply side, independently of the demand, may take on the system integrator's role and apply integrated production and business formats. On the supply side, the system integrator would normally be a main contractor but could also be a project developer or an architect. Obviously, the system integrator needs to integrate the *whole supply system*. The integration of the supply system is not only driven by *economic* arguments but also includes *organisational* and *social* aspects between firms and people (Bridge 2004, Bridge 2005). Analogously to the demand system integrator, the supply system integrator will apply *multi-project strategies* and establish *close inter-firm relations*. Those arrangements are aimed at operational and competitive *advantage* in relation to competitors (Nobeoka & Cusumano 1997). Project-independent *production strategies* such as platform strategies and modularity, have in some cases, particularly in housing, been demonstrated to be possible and beneficial (Vrijhoef & Voordijk 2004).

The idea of the system integrator is useful for the focal firm in the integrated supply chain. While it is ambiguous which firm type in the building supply chain is the focal firm, the idea needs to be specified per firm type. This research is primarily focused on the supply side rather than the demand side of the building supply chain, including the client organisation. Therefore, from this point on we will concentrate on the supply side issues of supply chain integration, and thus on the role of the supply system integrator.

3.3 Contributing topics to supply chain integration in the building industry

In addition to the thoughts above pointing in the direction of a concept of supply chain integration for the building industry, a number of additional topics have been previously raised in building literature, and which also contribute to the concept of supply chain integration directly or indirectly (Table 6).

Table 6: Overview of contributing topics to the concept of building supply chain integration

Topic	Contributing ideas to the concept of building supply chain integration	References
Integrated contracts and strategic procurement	Firms are unified in collective and often long-term contractual arrangements delivering an integrated product to the client	Cox & Townsend
Alliances and partnering	Firms team up as partners in integrated teams per project or for a long term	Bresnen & Marshall
Lean and agile construction	Increased effectiveness and efficiency of operations via intensified control and increased flexibility of production	Koskela, Howell, Ballard, Naim
Co-makership	Early and intensified involvement of subcontractors and suppliers in the engineering and manufacture	Vrijhoef, Iwashita
JIT logistics	Increased involvement of suppliers and transport firms delivering material to site lowering the inventory on site	Johnston, Pheng
Modularisation, standardisation and postponement	Modular approach to design, production and site installation enables standardisation and postponement	Van Randen, Dekker, Wolters
Prefabrication, offsite construction and industrialisation	Improved allocation and fit of production resources through a shift of activities off site	Koskela, Sarja, Warszaki

In their own right, these topics provide partial contributions to the concept of supply chain integration. The collection of the contributions and the references they make to building practice give initial shape and grounding to the concept of supply chain integration to be further developed in the course of the research.

3.3.1 Integrated contracts and strategic procurement

Application of strategic procurement approaches have led to tendering packages of project activities to clusters of firms for longer periods of time, e.g. via integrated PFI and DBFMOT kinds of contracts[1] (Ndekugri & Corbett 2004, Nicolini et al. 2001). These *integrated kinds of contracts* have generally improved the economic efficiency of the integrated design, build and facility management through cost minimisation over the whole life cycle. Integration of the supply chain via an integrated contract is a logical measure by the client to seize the opportunity and the incentive to implement life cycle performance-driven design solutions. This is a key element for success, yet the opportunities become constrained and the incentives weakened as projects progress and are handed over from one party to the next in the supply chain (Rintala 2004). The use of integrated contracts tends

1 PFI = Privately Financed Initiative types of contracts leave the financing of a project to the delivering party or parties. Generally the built facility remains the property of the delivering party or parties, and the client pays rent or a usage fee. DBFMOT = Design-Build-Finance-Maintain-Operate-Transfer types of contracts (or a shorter construct, e.g. DBFM contracts) are integrated contracts including transfer of responsibility of a number of the DBFMOT aspects to the supplying party or parties.

avoid this while contractual terms not only form a legal tie but are also a way in which knowledge and activity flow between organisations involved within a building project (Hall et al. 2000). New procurement methods have also been regarded as a tool to integrate the building supply chain, because they provide openness, trust, cooperation, benefit sharing, collective working routine, and fair allocation of risks (Benheim & Birchall 1999). Often such new procurement methods thrive on long-term perspectives and relational aspects (Cox 1996, Cox & Townsend 1998). Many of these clients create multi-project environments implying multi-project procurement and repetitive tendering arrangements with contractors (Blismas et al. 2004a). In this perspective contractors are observed as partners adding value to the client's business instead of merely accomplishing projects for the client (Bresnen & Marshall 2000a). In this respect integrated contracts and strategic procurement add to supply chain integration via long-term contractual arrangements delivering an integrated product to the client organisation, for a longer period in the product's life cycle. Often this kind of arrangements involves firms such as investors, project developers and contractors, rather than designers and suppliers. The client organisation is generally excluded from the actual supply chain after transferring most or all of the work and risk of the entire product delivery and facility management to the contracted parties.

3.3.2 Alliances and partnering

The concept of alliances is useful as a common starting point for discussions between a client and a main contractor on how to procure a specific alliances project, thus avoiding misinterpretations and failures (Yeung et al. 2007). In addition it has been advocated to target the supply chain downstream (subcontractors, material suppliers, etc.) as a means of effectively reducing overall building costs (Proverbs & Holt 2000). This has been referred to as 'downstream strategic alliances', involving subcontractors and suppliers early. This gives an opportunity to downstream participants to offer their expertise, which could potentially result in *cost and risk reduction*. Advanced building clients are moving towards the use of *framework agreements* and *multi-tier sourcing* techniques to rationalise the supplier base (Fernie et al. 2000). As an alternative route, partnering has been presented as a way of improving performance in building through the benefits it can bring to both clients, contractors and additional firms in such long-term arrangements (Bresnen & Marshall 2000a, Bresnen & Marshall 2000b). On the other hand, there is still considerable debate about the nature and merits of long-term partnering approaches in contrast to temporary and often more traditional procurement approaches (Bresnen & Marshall 2000c, Bresnen & Marshall 2002). The ideas of alliances and partnering particularly contribute to the concept of supply chain integration while the supply chain teams up as partners in integrated teams per project or for a long term contractually. However such arrangements often tend to tie partners together contractually, and not necessarily operationally.

3.3.3 Lean and agile construction

Since the early 1990s, lean production in manufacturing, via the concept of lean supply, has been a major source for the emergence of supply chain management (Lamming 1993, Womack et al. 1990, Womack & Jones 1994, Womack & Jones 1996). The same is true for lean construction and supply chain management in the building industry (Koskela 1993, Koskela et al. 2002). Lean construction initially emphasised *production control and planning*, e.g. collaborative planning with subcontractors and suppliers, through systems such as the last planner system (Ballard 2000). The logic of lean production as applied in manufacturing is based on the idea that supply chain management is more effective with *stable supply chains* and with *standardised or 'mass-customised' products*. However, both features are restrictive with respect to the required flexibility in traditional building. From this point of view, Naim et al. (Naim et al. 1999) suggest developing building supply chains following an 'agile' paradigm, i.e. using market knowledge and a virtual corporation to exploit profitable opportunities in a volatile market. In addition, Naim et al. (Naim et al. 1999) suggest the potential to develop '*leagile*' building supply chains applying standard components and a postponement strategy that can respond to changing customer requirements in an effective and efficient way. Lean and agile construction contribute to the effectiveness and efficiency of supply chain operations via increased involvement of subcontractors and suppliers in the planning and control of site construction particularly. This however does not necessarily imply long-term and strategic supply chain arrangements and remains relatively operational. Generally it lacks an economic perspective.

3.3.4 Co-makership

Like lean construction, co-makership originated in manufacturing (Broersma 1991, Merli 1991) and found its way to building (Vrijhoef 1998). The central idea is to *involve suppliers* as early and intensively as possible in the design engineering and manufacture of parts, as if they were 'external business units' of the manufacturer (Amaral et al. 2002, Bidault et al. 1998, Dowlatshahi 1999). In the building industry, the emergence of co-makership has been observed in the trend of *systems subcontractors* delivering and installing integrated parts of buildings directly on site (Iwashita 2001). In addition, major opportunities have been identified for the application of *systems sourcing* to promote such an approach to subcontractors as system suppliers in the development and manufacturing of building parts, and intensified deployment of the technological capabilities of suppliers (Gadde & Jellbo 2002). The idea of co-makership aligns well with supply chain integration in particular with regard to the early and intensified involvement of specialist subcontractors in the engineering and construction via rationalised sourcing methods. Co-makership is limited though to the relation between the main contractor and specialist subcontractors in the engineering and construction phases. This would be more effective if product development and design were included, as well as the experts involved in those activities, since design is often separated from construction.

3.3.5 Just-in-time logistics

Initial applications towards supply chain integration in the building industry have originally been in the field of logistics and materials control on the building site (Asplund & Danielson 1991, Johnston 1981, Wegelius-Lehtonen & Pahkala 1998). There has been a shift from the material handling and logistics occurring on site to the cooperation between suppliers and contractors to *improve the total flow of material* under dynamic circumstances (Voordijk 1999). Centralisation and integration of *transport and distribution* by traders of building materials have been among the major developments in the building supply chain. Mergers and acquisitions created big retail chains with their own central warehouses (Voordijk 2000). Most of these firms have centralised their *inventories*. This centralisation and the use of *ICT* resulted in lower inventories and more efficient transport. Notably, the introduction of *just-in-time logistics*, as part of the concept of lean construction, has had a great influence on shorter delivery times and lower inventories on building sites (Pheng & Chuan 2001). Increased involvement of suppliers and transport firms and more advanced logistics solutions such as delivering smaller badges more often to site and lowering the inventory on site can be regarded as general aspects of supply chain integration. However this does not necessarily include centralised logistics control in the supply chain.

3.3.6 Modularisation, standardisation and postponement

Modularisation of buildings has been a movement in the building industry since the 1960s, notably in the *open building* movement in housing (Dekker 1998, Van Randen 1990). Open building can be viewed as a *postponement* strategy. The delayed differentiation, product modularity and the separation of different levels of decision-making enable the decoupling of building parts within a modular system (Vrijhoef et al. 2002). Three types of *modularity* have been identified which can be applied to buildings with an impact on the organisation of the supply chain (Baldwin & Clark 2000): modular-in-production, modular-in-design and modular-in-use. *Modularity-in-production* rationalises a product into components and allows parts to be standardised and produced independently before assembly into the final system. *Modularity-in-design* goes a step further with an overall architecture and standard interface;, the modules can be designed independently, and mixed and matched to create a complete system. Finally, a product is *modular-in-itself* if consumers themselves can mix and match components to arrive at a functioning whole. Modularity has often been applied to houses and offices (Wolters 2000). In practice this has led to the application of modular units manufactured in a factory, including all installation and finishing. The assembly of the units is done on site. The connections between the modules are *standardised*. This simplifies the scheduling and the work on site and reduces overall building time. The rationalised approach to product and process eases the involvement of the supply chain, i.e. suppliers involved in the development and manufacture of standardised building parts. Modular approach to the design, production and site installation enables standardisation and postponement. Supply chain integration bridges the traditional separation between design and construction that prevents these issues to be used to their full extent.

3.3.7 Prefabrication, offsite construction and industrialisation

In effect, many of the above topics have led to the transfer of building activities off site as a result. Industrialisation, especially prefabrication, can be regarded as a structural means for *eliminating on-site activities* from the total production chain (Warszaswki 1990). Transferring activities off site has its implications, though. The structure and behaviour of the total process change. The process generally gets longer, the design required is more substantial, the error correction cycle gets longer, and the required dimensional accuracy usually needs to be higher (Kazi et al. 2007, Sarja 1998). The total process of industrialised building thus tends to become *more complex* and *vulnerable to variability* as compared to traditional on-site construction, even if the part of the process located on site becomes less complex (Koskela 2000). The dissemination of industrialised building, with its longer and more complex supply chain, has called for integrated supply chain arrangements across the building industry (Kazi et al. 2007). Improved quality, increased productivity and shielding of production resources are generally the effects of a shift of activities off site. But again supply chain integration needs to relieve the traditional separation between design and construction, ensuring the effects.

3.4 Reflection on key issues and contributions of this chapter to the thesis

This chapter has presented a first set of a priori constructs framing the further development of the concept of supply chain integration in the building industry, at which this thesis is aimed. The constructs have included indications of lessons the building industry could draw from manufacturing, previous thoughts pointing in the direction of supply chain integration. The ideas that have emerged in recent decades in building literature related to supply chain integration, offer direction to the concept to be developed in this thesis. This includes conceptual contributions as well as gaps and limitations identified offering further direction (Table 7).

Although the issues presented have provided initial grounding for the concept of supply chain integration, they have also indicated gaps in the current understanding of supply chains in the building industry and related topics, which the concept to be developed will need to address.

Table 7: Overview of constructs presented in this chapter, their conceptual contributions and gaps

Constructs	Conceptual contribution to the concept of supply chain integration	Gaps and limitations
Quasi firm	The supply chain is regarded as a single firm.	The idea has mainly applied to contractors and subcontractors, excluding firm types such as developers, designers etc. The idea is predominantly economic, underexposing aspects of production and organisation.
Supply chain management	The building supply chain is managed in an integrated way.	Generally the idea regards the main contractor as the focal firm, which is not necessarily so in building practice.
Systems integrator	The supply chain viewed as an coherent organisational system is led by the focal firm.	This idea needs to be specified per firm type in the building supply chain, since the position of the focal firm is ambiguous in building practice.
Integrated contracts and strategic procurement	The supply chain is contracted via long-term contractual arrangements to deliver the product including additional services before and after delivery.	Generally this applies to developers and contractors, not to designers and suppliers. The client organisation transfers risk and work to contracted parties, and is not involved in the product delivery.
Alliances and partnering	The supply chain teams up as partners in integrated teams per project or for a longer term contractually.	Generally the integration remains contractually, not operationally nor organisationally.
Lean and agile construction	Increased effectiveness and efficiency of operations via involvement of subcontractors and suppliers in production control and planning	This does not necessarily imply long-term arrangements and remains operational. Generally it lacks an economic perspective.
Co-makership	Early and intensified involvement of subcontractors and suppliers in the design, engineering and manufacture via systems sourcing.	This includes the contractor and specialists for engineering and construction. It lacks a view on product development and design, since those are often separated in building, and dedicated to separate designer.
JIT logistics	Increased involvement of suppliers and transport firms delivering smaller badges more often to site lowering the inventory on site.	Generally this is still based on traditional and individual logistics coordination, not necessarily centralised logistics control in the building supply chain.
Modularisation, standardisation and postponement	Modular approach to the design, production and site installation enables standardisation and postponement.	The traditional separation between design and construction prevents these issues to be used to their full extent.
Prefabrication, offsite construction and industrialisation	Improved quality, productivity and shielding of production resources by moving activities off site.	The traditional separation between design and construction prevents these issues to be used to their full extent.

4 Theoretical framework

Now that the concept of supply chain integration has been framed based on thoughts and topics existing in building practice and literature, this chapter will present a multiple theoretical framework based on general theory. The framework aims to theoretically explain and understand the organisation and functioning of supply chains from four theoretical perspectives of governance: economic governance, production management, inter-firm governance, and social governance. At the end of the chapter, the four perspectives are synthesised in a joint set of theoretical constructs. Together with the a priori constructs presented in the previous chapter this will be operationalised into an analytical framework for supply chain integration in the following chapter, which will be used for the empirical analysis and the synthesis in the following parts of the thesis.

4.1 Explaining the need for a multiple theoretical view on the supply chain and the selection of the four perspectives

Although a mono-theoretical perspective would be sufficient in order to understand sole aspects of the supply chain, a full understanding of the *various aspects*

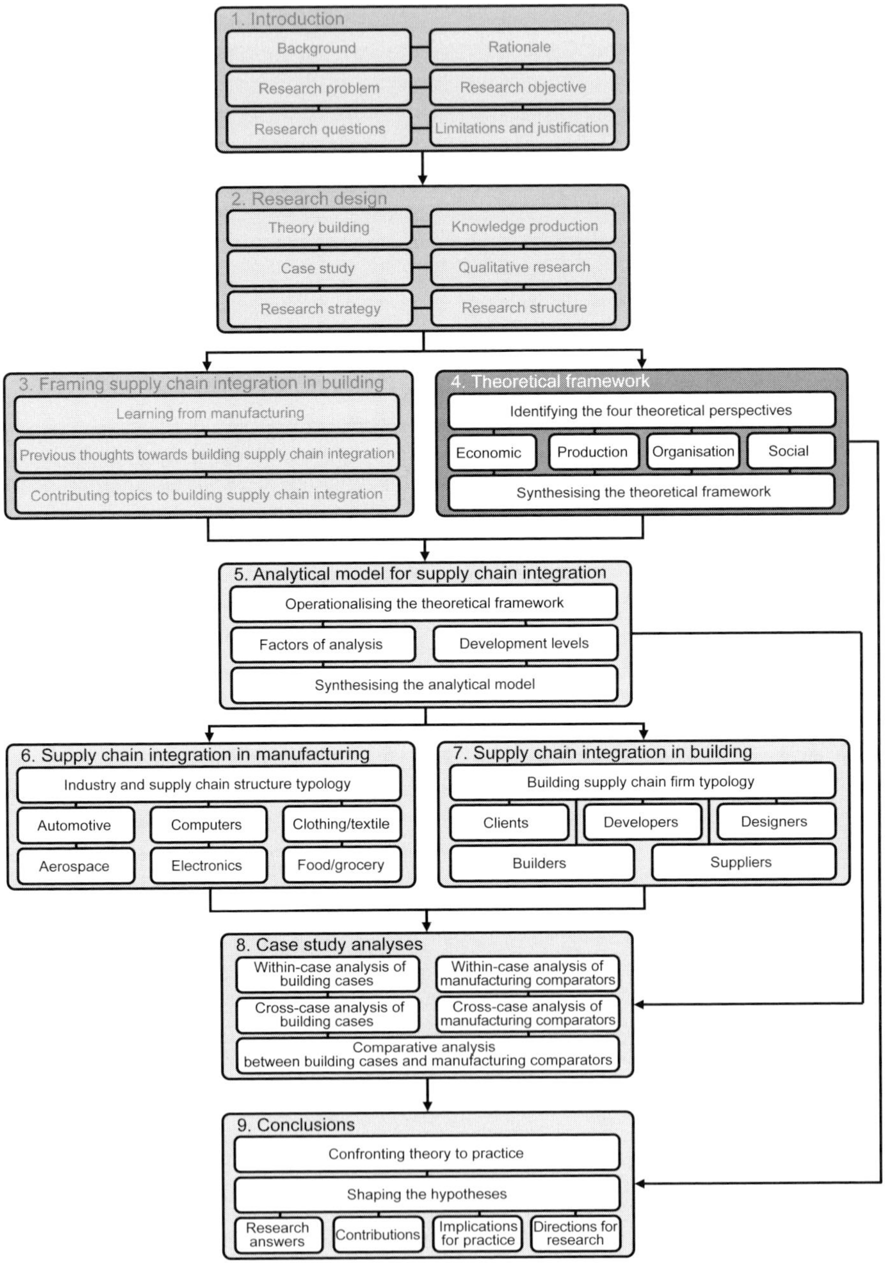
1. Introduction
Background
Rationale
Research problem
Research objective
Research questions
Limitations and justification
2. Research design
Theory building
Knowledge production
Case study
Qualitative research
Research strategy
Research structure
3. Framing supply chain integration in building
Learning from manufacturing
Previous thoughts towards building supply chain integration
Contributing topics to building supply chain integration
4. Theoretical framework
Identifying the four theoretical perspectives
Economic
Production
Organisation
Social
Synthesising the theoretical framework
5. Analytical model for supply chain integration
Operationalising the theoretical framework
Factors of analysis
Development levels
Synthesising the analytical model
6. Supply chain integration in manufacturing
Industry and supply chain structure typology
Automotive
Computers
Clothing/textile
Aerospace
Electronics
Food/grocery
7. Supply chain integration in building
Building supply chain firm typology
Clients
Developers
Designers
Builders
Suppliers
8. Case study analyses
Within-case analysis of building cases
Within-case analysis of manufacturing comparators
Cross-case analysis of building cases
Cross-case analysis of manufacturing comparators
Comparative analysis between building cases and manufacturing comparators
9. Conclusions
Confronting theory to practice
Shaping the hypotheses
Research answers
Contributions
Implications for practice
Directions for research

in the exchange between firms and humans within the supply chain requires multiple perspectives on the supply chain (Cousins et al. 2006). Several theories and concepts apply to the phenomena existing in the supply chain. These theories and concepts can be arranged in larger theoretical domains. Since this thesis focuses on the *integrated management and governance* of processes and activities through the supply chain corresponding topics apply. To this end, Porter's *value chain* model has been used (Porter 1985) (Figure 12). The value chain is aimed at the overall coordination of the firm as a system of activities aimed at maximisation of the firm's margin and the value delivered to the customer.

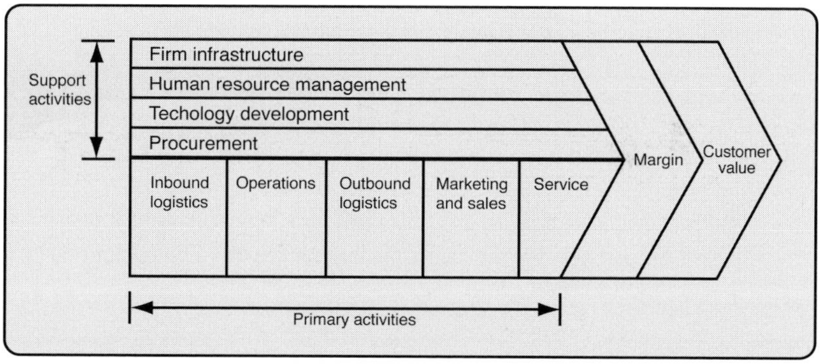

Figure 12: Porter's value chain indicating the perspectives on the supply chain. Adapted from Porter (1985)

In a supply chain perspective each firm delivers to the subsequent firm adding to the supply chain's collective output via its own set of activities. The activities include primary and support activities. The *support activities* cover the primary activities. In this thesis, all management and governance of the primary activities are considered to take place within the support activities. In addition, Porter (1985) distinguishes, per activity type, direct and indirect activities. *Direct activities* are 'directly involved in the value creation', i.e. business level. Indirect activities 'make it possible to perform the direct activities', i.e. corporate level. Supply chain integration focuses on the integrated management and governance of the extended delivery process throughout the supply chain, which implies an integrated approach to support activities that are directly connected to the primary activities, and not necessarily the integration of primary activities themselves. Therefore, the thesis will focus on the role of *direct support activities*, rather than indirect support activities (Table 8).

The four types of support activities point in the direction of four broader sets of theoretical topics, respectively firm infrastructure, i.e. organisational topics; human resource management, i.e. social topics; technology development, i.e. production topics; and procurement, i.e. economical topics (Figure 13). Following Porter (1985), the firm infrastructure refers to the 'overhead' of an organisation. In a supply chain perspective, and for the purpose of this thesis, this idea is broadened to the *organisation of the supply chain*. Human resource management refers to the personnel policy of an organisation to improve the capabilities of the personnel to contribute to the organisation's business objectives (Porter 1985). In a supply chain perspective, the *collective personnel in the supply chain* needs to be addressed. This adds the firm resources to the human resources. For the purpose of this thesis, the resource policy towards both supply chain firms and their personnel is bundled and widened in a *social perspective to the supply chain*. Technology development refers to the collection of technological capabilities to be able to develop and produce products (Porter 1985). In a supply chain perspective, for the purpose of this thesis, this idea is focussed to the *integrated production function of the supply chain*. Finally procurement refers to the economic transactions of a firm to purchase external inputs (Porter 1985). In a supply chain perspective, for the purpose of this thesis, this aspect is rephrased to the *transaction economics of the supply chain*.

Table 8: Support activities included in the thesis based on Porter's value chain model (Porter 1985)

Support activities	Activities included (direct)	Activities excluded (indirect)	Explanation
Firm infrastructure	• Planning • Quality management	• General management • Finance, accounting, • Legal, government affairs	Business departments directly adding to value creation are included. Corporate departments are excluded.
Human resource management	• Training, development • Improving skills and motivation	• Recruiting, hiring • Compensation, remuneration • Negotiation, assessment	Direct development of human resources in the value creation is included. Indirect development is excluded.
Technology development	• Improvement of product and process	• Research and development • Media research • Process equipment design • Servicing procedures	Direct management and improvement of value creation i.e. products and production. Indirect R&D are excluded.
Procurement	• Purchasing materials and parts	• Purchasing general supplies • Firm's cost calculations	Procurement of direct inputs for value creation i.e. production are included. Indirect inputs are excluded.

Previous attempts in building research and construction management research particularly have been undertaken to chart the field of building supply chain research (London & Kenley 2001, O'Brien et al. 2002). Following those attempts this thesis will address a multiple theoretical perspectives to the supply chain, and thus it applies a multiple theoretical framework. This framework is based on the *four theoretical directions* identified above. Following Eisenhardt (1989) a priori constructs will be identified in these theoretical directions to ground and direct the theoretical contribution of the thesis. Moreover, *gaps and limitations* in the existing theory may provide opportunities to enrich the theoretical contribution of the thesis.

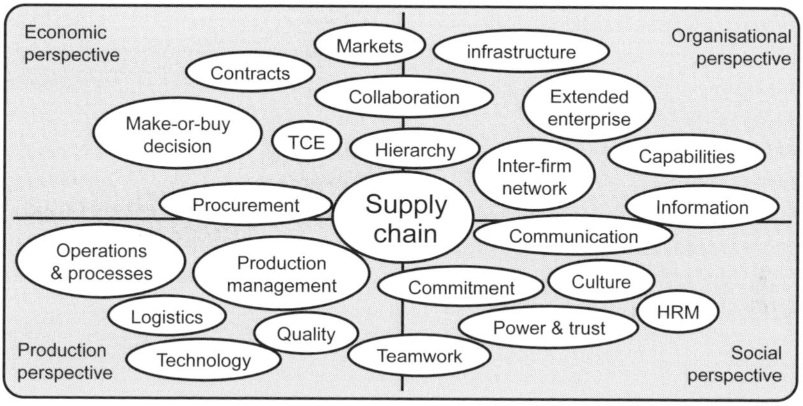

Figure 13: Four theoretical perspectives on the building supply chain. Adapted from Vrijhoef et al. (2003)

In each of the four theoretical perspectives the thesis will address a major theoretical theme to deduce the theoretical constructs used further in the thesis, and identify the gaps and limitation the concept of supply chain integration would need to resolve in to increase its applicability in the building industry. In the economic perspective, the focus will be on the theme of *transaction cost economics (TCE)*. In the production perspective the focus will be on the theme of *production management* related to supply chain management. In the organisational perspective, the focus will be on the theme of *inter-firm networks*. Finally, in the social perspective, the focus will be on the theme of *communication and commitment*, as part of the social interaction between firms and humans within the context of the supply chain.

4.2 Introducing the four theoretical perspectives on the supply chain

4.2.1 Economic perspective: transaction cost economics

The principles of TCE contribute to the supply chain in the manner of the choice of governance structure made by economising on the total sum of production and transaction costs. In this approach production costs are considered to be constant, thus leaving transaction costs as a main target for minimisation. *Transaction costs and their reduction* lie at the heart of the interest in supply chain management. Transaction costs are associated with search and information costs, bargaining and decision costs, policing and enforcement costs (Dahlman 1979). TCE provides an *explanation for the existence and structure* of firms within a supply chain (Hobbs 1996). The central logic of TCE is that when transaction costs are low, *contracting* is used (i.e. market), while *internalisation* will prevail for high transaction costs (i.e. hierarchy). Intermediate modes are often referred to as hybrid modes (Williamson 1991, Williamson & Winter 1991). Another intermediate form of governance is the *network* (Powell 1987, Powell 1990). TCE recognises that transactions do not occur without friction. Costs arise from the interaction between and within firms as *transaction costs*: information costs, negotiating costs and monitoring costs i.e. enforcement costs (Hobbs 1996). Human factors such as *opportunism* and *bounded rationality* influence the transaction costs. Transaction costs would be zero if humans were completely honest and possessed unbounded rationality. Alternatively *social interaction* and *integrated decision making* decrease opportunism and increase rationality (Moon & Kim 2005).

Transactions costs for a particular transaction depend on the three critical dimensions of transactions: asset specificity, uncertainty and frequency of transactions (Williamson 1985a, Williamson 1985b). *Asset specificity* is an important factor that characterises the governance mode of transactions (Williamson 1975). It refers to the situation in which a firm and selected suppliers need to engage in specific investments in order to make transactions possible. Because of opportunistic behaviour, resources will be spent on contractual and organisational safeguards, and indeed 'trading partners protect themselves from the hazards associated with exchange relationships' (Shelanski & Klein 1995). Loader (1997) concludes that *uncertainty* generally drives a supply chain towards vertical integration, depending on the extent of asset specificity and uncertainty regarding price levels. On the contrary, improvement of the levels of information among parties along the supply chain may relieve the uncertainty and in this way be a substitute for vertical integration (Loader 1997). A corresponding argument is applied to *frequency*. The greater the frequency of transactions the more likely firms will be to engage in vertical integration and reduce transaction costs that come with market searching and external contractual arrangements (Ekstrom et al. 2003). In addition, Milgrom and Roberts (Milgrom & Roberts 1990, 1992) add two other issues: *difficulty of performance measurement* and *connectedness to other transactions*. Both are particularly relevant from a supply chain viewpoint.

The aim of supply chain integration is to improve coordination and communication in the supply chain, and thereby reducing transaction costs (Hobbs 1996). Rather than treating each transaction separately, benefits are to be gained from organising clusters of related transactions within the supply chain (Williamson 2008). Following the logic of TCE, change in the transaction costs arising from the exchange of a product may lead to a change in the governance of the supply chain, depending on the degree of uncertainty, asset specificity and frequency. Generally high levels of uncertainty, asset specificity, and infrequency would lead to more formal types of supply chain integration, and possibly vertical integration, i.e. internalisation (Blois 1972, Díez-Vial 2007).

4.2.2 Production perspective: production management

In addition to TCE's focus on transaction costs, production management is aimed at minimising production costs. In terms of *lean production*, particularly any form of waste i.e. non value-adding activities existing in the production system of the firms within the supply chain will need to be reduced or removed. This view is compatible with the decomposition principle of production (Koskela 2000), where production is conceived as a series of decomposed tasks and semi-products leading to an integrated final product. Instead of minimising a specific type of *waste* associated with one activity type in production, the objective is rather to organise the entire production through the supply chain in such a way that the aggregate amount of waste is minimised.

However, the selection of an organisational form to organise the production is focused on the general objective of *value creation* for end customers, rather than just waste minimisation, which may be regarded as a sub-goal of production management (Galbraith 1995). Seeking a general explanation for the selection of an organisational form of production on the basis of minimising waste or production costs only is not justified. Instead, the supply chain, defined as 'the network of organisations that are involved, through upstream and downstream linkages, in the different processes and activities', must produce 'value in the form of products and services in the hands of the ultimate customer' (Christopher 1992) (Figure 14). Supply chain integration must therefore have a positive effect on the integrated delivery of end products and services in the hands of end customers' (Hobday et al. 2005).

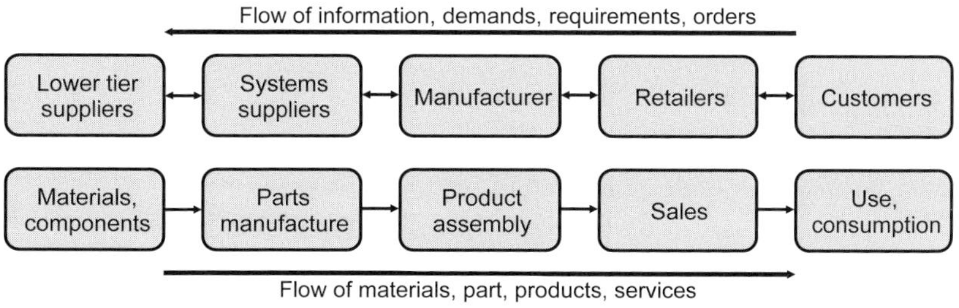

Figure 14: Generic configuration of a supply chain in manufacturing (Vrijhoef & Koskela 2000)

As a result, production management in essence becomes *supply chain management*. Christopher (1992) observes that 'supply chain management covers the flow of goods from supplier through manufacturing and distribution chains to the end user'. This means that the independent firms within the supply chain agree upon the way in which production and information flows are organised. Typical consequence of this kind of agreements is an integrated planning and organisation of activities and logistics among the group of firms involved (Bowersox & Closs 1996).

4.2.3 Organisational perspective: inter-firm networks

Contemporary industrial organisation and production strategies increasingly put emphasis on activities that are *external* to traditional organisational environments and on managing operations in an inter-firm network (Grandori 1999a, Grandori 1999b). This has caused a shift from the traditional enterprise of one organisation to an *extended enterprise*, virtual organisation or an "extraprise" shared amongst multiple organisations (Boardman & Clegg 2001, Cooper et al. 2003, Karlsson 2003, Stock et al. 2000). A similar model of industrial organisation is the modular production network (Sturgeon 2002, Sturgeon 2003). These network concepts imply *clusters of organisations* that work more intensely with each other than with other organisations within the industry (Ebers & Jarillo 1998). The organisations involved have developed *recurring ties* and *joint activities* when serving a particular market. A network can arise from a strategy of co-specialisation by which member firms carve out new and profitable product-market niches.

Distributed production activities and business functions within networks raise the issue of bundling and aligning *core competences* of individual firms as if they were one firm (Prahalad & Hamel 1990). Synergies can be realised by the combination of *complementary resources* of different firms. These *synergies* are possible in the areas of design quality, product development or minimising manufacturing lead times (Parkhe 1993, Teece 1992). The coordination of networks requires organisational *alignment* of the multiple strategies, as well as cultural alignment to manage people working along extended business processes (Bititci et al. 2003).

Networks imply a form of *governance* that is distinctly different from market and hierarchy (Sydow & Windeler 1998). Networks are deemed to be more flexible and complex than hierarchies (Powell 1987, Powell 1990). Compared with markets, networks provide potential for *flexible integration* and production control (Håkansson & Sharma 1996, Håkansson & Snehota 1995). The balance between *control and emergence* is important, in the sense that control tends to detract from innovation and flexibility, and emergence decreases predictability and manageability of operations (Choi et al. 2001, Dooley 1997). A major challenge is to devise a governance model that balances control and emergence leaving room for the 'adaptive, collective behaviour of the supply chain' (Surana et al. 2005).

Besides transfer of specialised know-how and efficient access to information, another condition promoting the rise of networks is *mutual trust*. In Ouchi's (1980) terms this represents a clan organisation. Objectives are achieved through *mutual understanding* and through *mediating* differences and disagreements. A *long-term* outlook encourages firms to search for ways of jointly accomplishing tasks, learning and promoting trust and stability. The shadow of the future in long-term relations decreases opportunistic behaviour (Axelrod 1984). Trust is thereby generated, which is, as Arrow (1974) noted, an 'efficient lubricant to economic exchange', such as in the integrated supply chain.

4.2.4 Social perspective: communication and commitment

Organisations have been described as 'the coordination of efforts of people working on a collaborative task broken down in a set of specialised activities. Coordination is then achieved through communication' (Taylor 1993). Next, this communication needs to lead to *commitment* to get the jointly coordinated activities accomplished. This idea is represented by the language/action perspective (Van Reijswoud 1996). Regarding the practical significance of this perspective, two directions were pinpointed by Winograd and Flores (1986). First, the process of *requesting, creating and monitoring commitments* can be facilitated by systems for constructing and coordinating conversation networks. Second, people can learn to *communicate for action* by improving their skills in understanding requests, and making promises and commitments.

In addition to the role of language/action, the building of *communities of practice* in business environments appear important for the development of *social capital* and improved business performance (Lesser & Storck 2001). Connections and relationships among firms and individuals, and a *culture of collaboration* are essential to foster mutuality, openness and knowledge sharing (Barratt 2004). However, particularly in supply chains, this depends on a complex pattern of *exchange power* that underpins the network of buyer-supplier relationships (Cousins & Menguc 2006, Cox 2001, Ireland & Webb 2007). It is the nature of exchange power that defines the *possibilities* and *interests* of firms and humans, and thus determines whether they can and will engage in relationships with others in the supply chain. In this respect, *trust* and *transparency* are key issues that need to be fostered, as are *socialisation* and *learning* (Kwon & Suh 2005, Sense & Clements 2006).

4.3 Applying the four theoretical perspectives to the building supply chain

4.3.1 Economic perspective: transaction cost economics

In an attempt to interpret building from an economic point of view, TCE can be viewed as one specific theoretical angle used for understanding building supply chains. When analysing the governance of activities in the building supply chain from a TCE perspective, the issues of asset specificity, uncertainty and frequency to transactions need to be balanced with the characteristics of building (Winch 2001, Winch 2006). In building projects, *asset specificity* tends to be low during the pre-contract phase, but high during the post-contract negotiations. The chances for *opportunistic behaviour* are limited before the contract is signed. Ties among firms within the industry network are typically weak, and the client can choose relatively randomly from many developers, architects and contractors. After selection and contract award, the situation is reversed. A contract is signed with a small number of parties to whom the client is legally firmly bound under strict regulations. In addition to this, and because of the one-off nature of most contracts, in practice, the compliance with the details of the contract, the brief and cost levels after the contract is signed are often a concern. Moreover, since subcontractors and suppliers are mostly involved after the initial contract, this often implies deviations and usually an increase of previously estimated cost levels (Chang 2006).

Several types of *uncertainties* occur during the building process (Winch 1989). In traditional building, task uncertainties are caused by the fact that each project requires new design and production solutions, but expertise transfer is generally limited or not systematically organised. Weather conditions and geological aspects cause natural uncertainties. Organisational uncertainties are tensions in the temporary project coalitions. Contracting uncertainty is related to the uncertainty of the cost estimation and the relatively high share of a project in the company's total turnover. In addition, according to Lansley (1994), judging the performance of the service offered by the subcontractor represents an uncertainty, which determines the need for either market or hierarchical governance in building supply chains. In most cases in building, levels of uncertainty and performance ambiguity are relatively high, and objectives are often incompatible, making a hierarchical approach seem appropriate.

However, in the building industry, the *frequency* of transactions between parties is relatively low because of ever-changing project coalitions and the use of market-based bidding procedures during the selection. The temporary character of relations may also stimulate opportunistic behaviour of firms, who may try to obtain as much benefit as possible from contracts before the project's end.

According to TCE, the characteristics of asset specificity, uncertainty and frequency in building would theoretically lead to integration of building activities within a *hybrid or hierarchical governance* structure. However, subcontracting and outsourcing are common practices in building. Application of hybrid or hierarchical modes of governance would be the rational response to uncertainty, complexity and post-contract, bilateral monopoly situations. It would economise on bounded rationality and opportunism, ease the transfer of expertise and facilitate systematic feedback, as well as solving coordination problems with design attributes through *centralised co-ordination* mechanisms (Milgrom & Roberts 1990, Milgrom & Roberts 1992).

Alternatively, Winch (2001) points to the fact that the deployment of third parties, i.e. *trilateral governance*, a term previously coined by Williamson (1975) for *occasional transactions*, is a distinctive feature of transaction governance in many project-oriented industries such as the building industry. Third parties, also known as 'control parties' often perform a central role in managing the building supply chain (Winch & Champagnac 1995). Third parties act as a go-between in monitoring and controlling compliance with agreements, safeguarding the performance and flexibility of networks of firms, and forming an important part of the social capital that supports the organisation and functioning of supply chains (Nooteboom 1999).

4.3.2 Production perspective: production management

Production management in building has previously been called to address the value, transformation and flow aspects of building in an integrated manner (Koskela 2000). Based on this view, three fields of management can be defined in building: project management, process management and value management (Bertelsen & Koskela 2002). The traditional *project management* creates and maintains the relations between the *value* as defined in the design and the brief, the operations and the associated production capacities, and the materials, personnel and equipment needed for the delivery of the project to be performed by the contracted parties (Turner 2006).

The *process management* takes the role of coordinating the *production* process, including the information, materials, people and equipment needed. In building, it is a management type that to a great extent is informal and conducted on site, thus 'management by walking around'. In the end, the process delivers the actual value to the client, which also needs to be managed. In other words, value management ensures that the process generates the value that the client wants (Langford et al. 2003). As most of the product value is defined through the design, the *value management* during the building process mainly has to do with the monitoring of process-related values such as lead time, cost levels, quality control, and ultimately customer satisfaction. In order to be successful, the production management function must address multiple aspects, including and notably the project, the process and the value (Kelly & Male 2001).

4.3.3 Organisational perspective: inter-firm networks

In building projects, relations between firms are typically maintained for the *duration of the project*. This challenges the transfer of expertise and systematic feedback on planning, design, construction and maintenance between parties. A more *centralised and repetitive governance* in the building industry has therefore been proposed as a logical path to decrease costs and time as well as increase quality and value delivered to the client and the user (Voordijk et al. 2000). Centralised forms of organising building supply chains, however, cannot completely be explained in terms of market and hierarchy. Co-operation and integration between supplying, constructing, and designing parties in networks make it possible to deliver an *integrated product* with quality guarantees to the market. In this view, instead of bounded rationality, know-how is transferred between firms for *joint product development*. Opportunistic behaviour would then be replaced by *mutual trust*.

Mutual trust is necessary for an open dialogue and effective knowledge sharing. Extending best practices downstream along the supply chain thereby enables further *reductions of transaction costs*. Cooperative learning and knowledge sharing in *alliance* or *partnering* kinds of cooperation can reduce cost levels between organisations even further (Love et al. 2002c). In addition, those kinds of cooperation support inter-organisational *learning* and improve total *quality* beyond the project horizon (Love et al. 2002b).

On an industry scale, Dubois and Gadde (2002) distinguish *tight couplings* within projects and *loose couplings* in the permanent network of the industry represented as a 'loosely coupled system'. The patterns of couplings influence productivity, innovation, and the behaviour of firms. In terms of organisational behaviour, cultural and human issues such as trust and learning have been indicated as having major implications for building supply chains e.g. (Love et al. 2002b). As a result, the network approach is able to improve not only the *performance* of supply chains, but also the *socio-organisational basis* of the inter-firm relationships within the supply chain (Pryke 2004, Pryke 2005).

4.3.4 Social perspective: communication and commitment

Particularly in less strictly organised supply chains such as in building, communication and commitment are key elements in *controlling* the supply chain. In building, supply chains form *temporary organisations*, where people from different organisations must collaborate and coordinate their tasks. Independent of the question whether particular work has been procured from the market or produced internally, the *coherence* of the communication and commitments through the building supply chain are essential. The language/action perspective has shown to be advantageous to the interpretation and execution of orders, and hence the business performance of the building supply chain (Vrijhoef et al. 2001). Therefore, understanding and *systematisation of the communication* between individuals and organisations in the supply chain are crucial. This is particularly the case in the less stable production environment and temporary production situation in building projects. The ability to flexibly establish and foster a *culture of collaboration* among individually operating firms and individuals in building projects is a key issue. This must increase the compatibility, trust and commitment which are generally lacking at the start of building projects but are necessary in order for collaborative strategies in the building supply chain to be successful (Crespin-Mazet & Ghauri 2007).

4.4 Synthesising the theoretical framework

In this chapter, *four theoretical perspectives* on the supply chain have been introduced: economic, production, organisation and social. The four perspectives are complementary and constructive in some respects as well as somewhat competing and incomplete with respect to achieving the thesis' aim of understanding the building supply chain, and the conceptualisation of supply chain integration for the building industry.

TCE as a governance system for the building supply chain is predominantly aimed at *economic* aspects of transactions without addressing additional aspects of production and organisation. From the *production* management perspective, not only transaction costs but production costs and added value of production are relevant, too. The production management perspective shows that an organisational form of production such as the supply chain must be aimed at the general objective of value creation through effective product delivery to customers, and must not merely be based on an economic argument. Both TCE and production management, particularly address bilateral inter-firm business relationships and do not include the complexity and additional aspects of wider, inter-firm networks. From the *organisational* perspective of inter-firm networks, alternative formats of industrial organisation are suggested, increasing the opportunities for lower transaction costs and increased production performance. In that sense, the perspective of inter-firm networks challenges both the TCE perspective (i.e. minimising transaction costs is not always the optimum), and the production management perspective (i.e. added value changes with different organisational forms of production). Finally, from the *social* perspective of communication and commitment, the network approach as a static alternative governance structure (besides hierarchy and market) must be extended with the dynamics of inter-human and inter-organisational communication and interactions, including the social and cultural issues of inter-human and inter-organisational collaboration.

In addition to providing understanding of the building supply chain, the four theoretical perspectives imply prediction of the building supply chain and provide initial meaning and grounding to the concept of supply chain integration in the building industry (Table 9). In the following chapter, the theoretical framework will be operationalised into an analytical model which will be used for the empirical research, and the further theory building in this research.

Table 9: Theoretical framework for supply chain integration in building

Theoretical perspective	Central issue	Main aim in the supply chain	Meaning for supply chain integration
Economic perspective: Transaction cost economics	Transactions between buyer and seller, characterised by asset specificity, uncertainty and frequency	Minimise transaction costs. Optimise governance structure determined by lowest transaction costs through the supply chain	Striking the right balance between market (purchasing) and hierarchy (internalisation) depending on the nature of transaction costs
Production perspective: Production management	Production constituting three different concepts: transformation, flow, value generation	Get the product produced; minimise waste; maximise value, e.g. reduce variability with all firms in the supply chain	Aligning production, optimising transformations and operations, ensuring flow, maximising value to the end customer
Organisational perspective: Inter-firm networks	Networks of actors interlinked by inter-firm relationships	Maximise value by collaboration, information exchange and decreasing opportunistic behaviour among firms and people in the supply chain	Organising for interaction and exchange by creating or modifying networks aimed at effective collaboration
Social perspective: Communication and commitment	Conversation; sending and receiving orders between individuals and organisations	Avoid breakdowns in conversation and commitment, create trust and transparency; foster learning and socialisation among firms and people in the supply chain	Organising for communication leading to commitment in the network of firms and people

4.5 Reflection on key issues and contributions of this chapter to the thesis

This chapter has presented a multiple theoretical framework. The literature study of the four theoretical perspectives leading to the framework has shown that no single theory appears sufficiently comprehensive to fully explain the multi-faceted phenomenon of the building supply chain. None of the theories as such represents a complete and useful conceptualisation for supply chain integration in the building industry. Instead, the theories show gaps and limitations in their application, particularly related to certain characteristics of building (Table 10).

However, in addition to the a priori constructs framing the concept of supply chain integration in a building context in the previous chapter, the generic constructs concealed within the theories presented in this chapter provide ingredients and further grounding for the development of the concept of supply chain integration for the building industry throughout the following chapters of the thesis.

Table 10: Overview of constructs presented in this chapter, their conceptual contributions, and gaps

Constructs	Conceptual contribution to the concept of supply chain integration	Gaps and limitations
Economic perspective: Transaction cost economics	Balancing rationally between market (purchasing) and hierarchy (internalisation) depending on the nature of transaction costs.	Irrationalities of balancing between the two, and relatively informal and ambiguous approaches to establishing transactions in building practice specifically need to be observed.
Production perspective: Production management	Aligning production, optimising transformations and operations, ensuring flow, maximising value to the end customer	The fragmentation of the delivery process to the end customer, particularly the separation between design and build obstruct flow and optimisation.
Organisational perspective: Inter-firm networks	Organising for interaction and exchange by creating or modifying networks aimed at effective collaboration	The industry fragmentation of many relatively small and separate firms hinders effective network forming.
Social perspective: Communication and commitment	Organising for communication leading to commitment in the network of firms and humans	Predominant project focus and existence of short-term inter-firm relations in building practice hinder communication and commitment.

AIRBU
AN EADS COMPANY
MSN 078
MSN 078

5 Analytical model for supply chain integration

To conclude the theoretical basis of this thesis, the a priori constructs identified in the conceptual and theoretical framework presented in the previous two chapters will be synthesised into an analytical model for supply chain integration. The model represents an operationalisation of the literature explored into a framework of ten factors of analysis constituting the conceptualisation of supply chain integration in this thesis. In addition, four development levels of supply chain integration are identified and added to the factors of analysis. The resulting analytical model will be used as a framework for the empirical analysis and as a basis for the further theory development throughout the remainder of this thesis.

5.1 Operationalising the theoretical framework into factors of analysis

The a priori constructs provided by the previous thoughts and topics from building literature as well as the theoretical framework presented in the previous chapters give *initial grounding* and direction for the development of a concept of supply chain integration for the building industry. In this chapter the constructs are

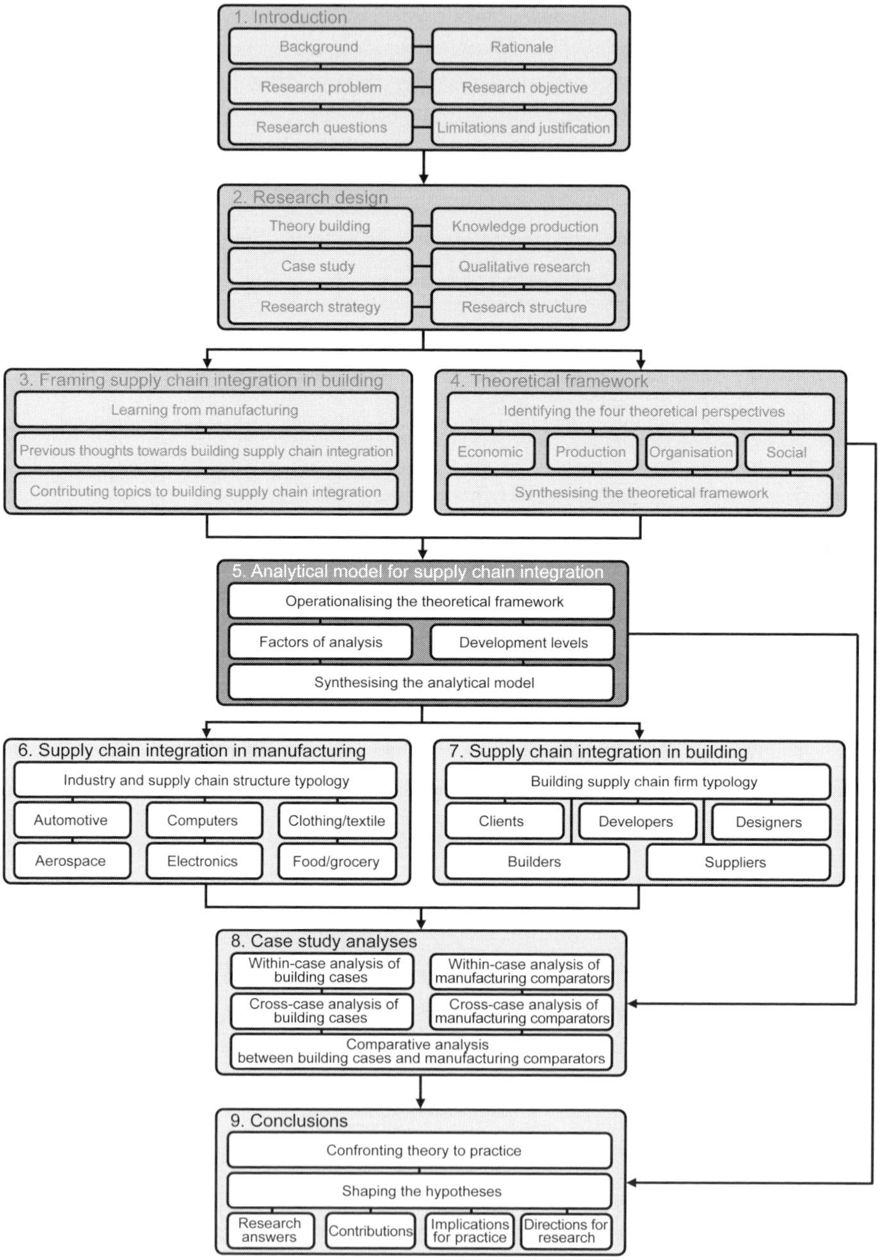
1. Introduction
Background
Rationale
Research problem
Research objective
Research questions
Limitations and justification
2. Research design
Theory building
Knowledge production
Case study
Qualitative research
Research strategy
Research structure
3. Framing supply chain integration in building
Learning from manufacturing
Previous thoughts towards building supply chain integration
Contributing topics to building supply chain integration
4. Theoretical framework
Identifying the four theoretical perspectives
Economic
Production
Organisation
Social
Synthesising the theoretical framework
5. Analytical model for supply chain integration
Operationalising the theoretical framework
Factors of analysis
Development levels
Synthesising the analytical model
6. Supply chain integration in manufacturing
Industry and supply chain structure typology
Automotive
Computers
Clothing/textile
Aerospace
Electronics
Food/grocery
7. Supply chain integration in building
Building supply chain firm typology
Clients
Developers
Designers
Builders
Suppliers
8. Case study analyses
Within-case analysis of building cases
Within-case analysis of manufacturing comparators
Cross-case analysis of building cases
Cross-case analysis of manufacturing comparators
Comparative analysis between building cases and manufacturing comparators
9. Conclusions
Confronting theory to practice
Shaping the hypotheses
Research answers
Contributions
Implications for practice
Directions for research

reviewed and reordered to achieve a more meaningful and useful operationalisation of supply chain integration particularly for the empirical investigation and further theory development of the thesis. The aim is that the operationalisation will bundle the a priori constructs in a focused and applicable framework that provides a better connection between theory and practice. This is achieved by the translation of the conceptual and theoretical constructs of the previous two chapters into more practical constructs that are more embedded in practice and the real world of firms operating in the supply chain. This has resulted in the set of corresponding directions for operationalisation given below, respectively dealing with economic governance, production management, inter-firm governance and social governance of integrated supply chains (Table 11). The directions have been broken down into major issues representing the *factors of analysis* used in the remainder of the thesis. Following the thesis' focus on the management and governance of the extended delivery process through the supply chain, corresponding topics of management and governance apply to the factors of analysis, rather than topics related to corporate finance or psychological aspects of organisations, for instance.

The aim of the *economic governance* of the integrated supply chain based on transaction cost economics is to minimise inter-firm transaction costs and ideally the integrated cost level of the collective activities throughout the supply chain. The economic governance is based on market, network or hierarchical relations. This gives rise to the strategic issue of whether to integrate business activities within one's own organisation (1), or to source them from other firms in the supply chain via strategic and partnering types of outsourcing and collaboration (2). Second, within the supply chain, production management is essential to control and align the total production across corporate boundaries, and to improve the performance of the collective production system of the supply chain towards delivering the end product to the end customer. Production management particularly aims at the overall coordination and integration of all operations and processes across the supply chain (3), including planning and logistics (4), and quality management of operations and processes (5). Third, the production operations and processes require substantial cross-corporate coordination, i.e. *inter-firm governance* between the firms in the supply chain. The interaction between the firms in the supply chain needs to maximise value for the end customer as well as for the firms themselves. To this end, the governance of the inter-firm relations particularly needs to address the issue of information exchange (6). In addition towards the end market, the issues of product development (7) and a joint market approach (8) geared towards the end customer are essential. Fourth, and to conclude, sustaining and supporting the level of communication and commitment among the firms as well as the individuals in the supply chain are key issues of the *social governance* of the integrated supply chain as an extended enterprise. Therefore, collectivism among the firms as well as individuals in the supply chain are essential and promoted via cultural alignment of the firms (9) and corresponding human resource management aimed at increased collective and synchronised action (10). These ten factors of analysis,

Table 11: Directions for the operationalisation of the theoretical framework

Theoretical perspectives	Conceptual constructs (building) (ch3)	Theoretical constructs (generic) (ch4)	Directions for operationalisation	Factors of analysis
Economic perspective: Transaction cost economics	The supply chain observed as a quasi-firm or extended enterprise, implying central control by the focal firm, via partnering and strategic arrangement including long-term contractual arrangements.	Minimising transaction costs requires economic governance of the supply chain aimed at lowest costs, particularly by balancing between market (contract) and hierarchy (vertical integration).	Economic governance of the integrated supply chain requires focal firm in the supply chain to consider internalisation of activities versus collaboration and outsourcing.	1. Integration of business activities in the focal firm 2. Partner sourcing and collaboration strategies with partner firms in the supply chain
Production perspective: Production management	Integrated approaches to the supply chain promote 'lean' and 'agile' operations and 'industrial' processes, prefabrication and offsite construction, particularly affecting logistics and quality.	Aligning production processes and operations ensuring flow aimed at maximising value to the end customer, minimising waste, reducing variability with all firms in the supply chain.	Production management of the integrated supply chain needs integrated approach to processes and operations, planning and logistics and quality management.	1. Integration of operations and processes 2. Planning and logistics control of operations 3. Integrated quality management of processes
Organisational perspective: Inter-firm networks	Both from demand and supply side the focal firm acting as systems integrator of the inter-firm network, applying modular approaches and standardisation with an impact on exchanging product and market information.	Organising for effective collaboration and collective action aimed at higher levels of collectivism and exchange towards the end customer of the supply chain.	Inter-firm governance of the integrated supply chain must lead to collective action and information exchange, joint product development and market approaches, marketing.	1. Information exchange between firms in the supply chain 2. Integrated product development of firms 3. Joint market approach to the end market
Social perspective: Communication and commitment	Early and intensified involvement of firms and their staff in building projects promotes a collective culture and working.	Supporting and sustaining conversation and commitment, among firms and people in the supply chain, aimed at collective culture and collaboration between firms and teams of individuals.	Social governance of the integrated supply chain promoting collectivism, commitment and cooperation between firms and individuals through cultural alignment and teamwork.	1. Cultural alignment between firm in the supply chain 2. Human resource management of individuals in the supply chain

together with the *development levels* per factor introduced later in this chapter, will serve as a model for the analyses of applications of supply chain integration in manufacturing and in the building industry in the following chapters.

5.1.1 Economic governance of the integrated supply chain

1 Integration of business activities

A first issue the focal firm needs to consider is the integration of business activities and the subsequent approach to the supply chain. The level of integration of business activities in the supply chain is essential. In the case of supply chain integration, the focal firm has integrated *internal business activities* and possibly *external business activities* as well. This depends on the *make-or-buy decision* of the focal firm per business activity (Bridge 2004, Bridge 2005). Essentially, this represents the choice that a focal firm has to either produce in-house or outsource activities externally. The focal firm needs to balance whether or not to take financial control or *ownership* over activities (Mahoney 1992). Financial ownership leads to *vertical integration*, which may at a certain level cease to be beneficial because of negative side effects of inflexibility (Buzzell 1983). Therefore, firms need to have a clear view on when to vertically integrate and when not to, and to what extent (Stuckey & White 1993). However, with inter-firm relations tending to become more complex over the past decade, *hybrid forms* have also emerged, e.g. groups of co-producing firms working under a shared brand (Klein 2005). Various strategies for vertical integration are possible (Harrigan 1983), such as *quasi-integration* (Blois 1972), implying a lower level of integration. This can be useful, depending, for instance, on the impact of *asset specificity* to the business of firms (Joskow 1988), or on the external sources of *innovation* for a firm (Robertson & Langlois 1995).

Factor of analysis 1: The focal firm has integrated business activities internally, and additionally controlling external business activities as well.

2 Partner sourcing and collaboration strategies

As a consequence of the approach taken to the integration of certain business activities, for additional activities and resources, the focal firm may enter into collaboration or sourcing strategies with other firms in the supply chain, e.g. establishing *strategic agreements* and contracts with partners' firms in the supply chain, typically with a *long-term* scope. Partner sourcing is essential in finding the right partner firms contributing value to the supply chain (Parfitt 2003). However, the value of *resources* procured from partner firms must not only be viewed from an organisational perspective, but must also be estimated in the eyes of the client (Santema & Van de Rijt 2005). With the partner firms, *strategic collaboration* may be established (Bennett & Peace 2006), and thereby shifting *procurement* and *purchasing* practices to a strategic level (Kenley et al. 2000). Strategic collaboration in general takes a long-term perspective to contractual relationships (Cooper et al. 1996, Haksever & Pickering 1996). Collaborative contractual relationships imply *non-adversarial bargaining* and *relational contraction* (Crook & Combs 2007). Once established, strategic collaboration enables adoption of advanced financial techniques among collaborating firms, such as *risk-reward schemes* and *risk and revenue sharing* (Giannoccaro & Pontrandolfo 2004, Lockamy & Smith 2000).

Factor of analysis 2: The focal firm has established strategic agreements and long term contracts with key partners in the supply chain.

5.1.2 Production management of the integrated supply chain

3 Integration of operations and processes

For supply chain integration to be effective, the focal firm and the partner firms in the supply chain will need to integrate their activities into an effective and efficient whole. Processes and operations across the corporate boundaries throughout the supply chain will thus need to be organised and integrated in order to strive for a seamless production *flow*. Process integration is a key aspect of supply chain integration (Trkman et al. 2007). The flow of production particularly refers to flows of *materials and operations* (Childerhouse et al. 2003b, Towill & Childerhouse 2006). The material is 'pulled' through the process (Ballard 1998, Tommelein 1998). *Lean* production as well as lean construction imply pull and flow and intensified co-ordination of the supply base, particularly to resolve waste and increase *efficiency* (Koskela et al. 2002). This is the same for agile production; however the latter is mainly focused on increased *effectiveness* and flexibility (Christopher & Towill 2000). However, particularly in a building context, the ability and the effects of applying certain types of production control is dependent on the type of project, varying from unique to standardised, and the type and scale of work, from small scale to large scale (Melles & Wamelink 1993).

Factor of analysis 3: Firms in the supply chain have connected or integrated their processes and operations to achieve flow, and increase the effectiveness and efficiency of the integrated delivery process.

4 Planning and logistics control

Related to the integration of processes and operations is the necessity for firms in the integrated supply chain to work together interactively in joint planning and logistics systems. Logistics is a key element of supply chain management (Bowersox et al. 2002, Lummus et al. 2001). In particular, *just-in-time* logistics is a part of integrated supply chains (Green & Inman 2005, Olhager 2002). This kind of logistics relies heavily on advanced forms of *inventory* management (Chandra & Kumar 2001, Lee & Billington 1992). The use of advanced *ICT* systems for joint planning are also essential enablers (Park 2005, Stevenson et al. 2005). Often, planning and logistics are outsourced to logistics service providers (Persson & Virum 2001, Rae-Smith & Ellinger 2002).

Factor of analysis 4: Firms in the supply chain jointly plan operations and processes, and organise and manage logistics.

5 Quality management

To conclude, a third important factor of production management in an integrated supply chain is quality management, and specifically an integrated approach to quality. Firms involved in an integrated supply chain, such as many in the manufacturing sector, typically apply *total quality* management across organisational boundaries and their collective production systems (Levy et al. 1995, Love et al. 2004a). Total quality management as described by Deming's seminal work (Deming 1986) has become a cornerstone of supply chain integration. Generally, the focal firm will take the lead in establishing a total quality approach to which the other firms will have to contribute. In addition, the *lean* and *six sigma* philosophies in particular have advocated continual review and improvement of quality (Harry & Schroeder 1999, Womack et al. 1990). In Japanese manufacturing and lean production, such a quality approach of *continuous improvement* has often been referred to as 'kaizen' (Imai 1986). Continuous improvement programmes ensure improvements to technology, materials, components, products, processes and services. In particular, relationship stability and integration are essential to ensuring consistently high quality and continuous improvement (Flynn & Flynn 2005, Fynes et al. 2005b, Lai et al. 2005). In a building context, however, in order to achieve continuous improvement effectively, supply chain firms need relatively *dynamic* quality systems that are externally orientated and flexible. Apparently, a combination of formal systems and relational aspects is essential to achieve high quality across the supply chain in the project environment of the building industry (Barrett 2000).

Factor of analysis 5: Firms in the supply chain apply total quality management to processes across corporate boundaries.

5.1.3 Inter-organisational governance of the integrated supply chain

6 Information exchange

In order to support an integrated approach to production management factors, firms in integrated supply chains must increase levels of *information sharing* and *shared knowledge* between themselves and other firms in and beyond the supply chain. The aim is to collectively and constantly be informed about the status of the supply chain, and to be able to mobilise the information and knowledge both within and outside the integrated supply chain needed for existing and new business. For that purpose, firms either rely on joint information *systems* (Holland 1995), or they foster the *flow* of knowledge and information between firms (Childerhouse et al. 2003a, Dimitriadis & Koh 2005, Egbu & Tong 2003).

Factor of analysis 6: Firms in the supply chain share information and knowledge in a joint information system.

7 Product development and design

Firms in the integrated supply chain particularly share information and knowledge as part of joint product development and design. The firms therefore pool resources along the supply chain, e.g. applying techniques such as *design for manufacture* to ensure the connection between design and production capability (Sehdev et al. 1995). In addition, the technological expertise of suppliers is often mobilised during the design by applying the concept of *early supplier involvement* (Amaral et al. 2002, Dowlatshahi 1998, Dowlatshahi 1999). Analogously, in a building context, early involvement of contractors has been found to be effective and efficient, particularly in case of continuity of work (Hughes et al. 2006). Product *modularisation* is often applied to enable the distributed production of parts or components of the final product (Lau & Yam 2005, Mukhopadhyay & Setoputro 2005). Product modularisation and distributed production are supported by application of *platform strategies* for developing a large set of interchangeable parts and components, and hence the ability to assemble different product types (Huang et al. 2005, Koufteros et al. 2002, Muffatto 1999, Xie et al. 2003). Distributed production may be organised in modular production networks of firms jointly manufacturing all parts and components for a final product in different locations (Sturgeon 2002, Sturgeon 2003).

Factor of analysis 7: Firms in the supply chain develop and design products jointly and rationally applying product and production modularisation.

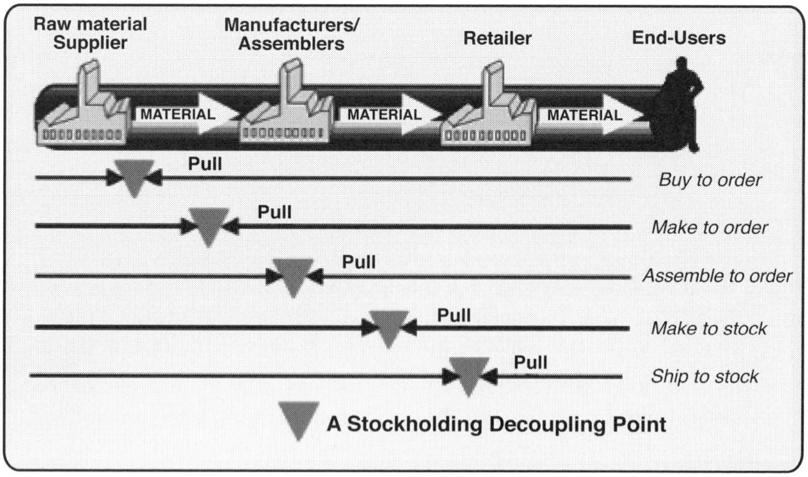

Figure 15: Decoupling point dictating the market strategy in the supply chain (Hoekstra & Romme 1992)

8 Market approach and marketing

Further to the product development and design, firms in an integrated supply chain are subsequently engaged in developing a joint market approach, via *collaborative marketing* strategies, e.g. joint venture, marketing channelling or co-development. Typically this process is led by the focal firm who will involve suppliers but particularly resellers and other firms in the sales channel to *co-develop* and commit to a coherent market approach that is generally fuelled and centrally controlled by the focal firm (Crespin-Mazet & Ghauri 2007, Gundlach et al. 2006, Tsang 2000). The *decoupling point* has played an important role, separating the part of the supply chain oriented towards customer orders from the part of the supply chain based on planning (Hoekstra & Romme 1992). The position of the decoupling point dictates the product market approach from 'buy to order', i.e. customisation, to 'ship to stock', i.e. mass production (Figure 15).

Associated with the decoupling point is the issue of *postponement,* i.e. late configuration. The postponement concept has been identified as a major factor influencing the market strategy of supply chains (Skipworth & Harrison 2006, Su et al. 2005, Van Hoek 1999). The principle of postponement can be subdivided into three generic types: form, time and place postponement (Bowersox et al. 2002, Bowersox & Closs 1996): 1) Form postponement entails delaying activities that determine the form and function of products in the chain until customer orders have been received. 2) Time postponement means delaying the forward movement of goods until customer orders have been received. 3) Place postponement refers to the positioning of inventories upstream in centralised manufacturing or distribution operations, to postpone the forward or downstream movement of goods.

Factor of analysis 8: Firms in the supply chain are engaged in collaborative market approaches with the aim of delivering integrated products to the end market effectively and efficiently.

5.1.4 Social governance of the integrated supply chain

9 Cultural alignment

Firms in an integrated supply chain implicitly or explicitly align their organisational cultures, and make the supply chain appear to be a 'single firm', i.e. promoting increased *collectivism* in the supply chain (Ryu et al. 2006). Cultural alignment has been referred to as aligning 'cultural capabilities' among different organisations. In this respect, organisational culture has been associated to the 'glue holding together the diverse values, beliefs, assumptions, understandings and norms that members of an organisation share, and the visible patterns of behaviour that an organisation manifests in the marketplace' (Gattorna 1998). In the supply chain, the different firms shift and shape their respective cultures to ensure that individual strengths are capable of supporting joint strategic objectives, and propagating a joint culture that is understandable for all employees in the supply chain and will govern their relationships (McAfee et al. 2002). This leads to higher levels of *cultural consistency* and assures effective communication, trust and interdependence among the firms and people in the supply chain. This process of *homogenisation of cultures* i.e. cultural alignment has been associated with enhanced supply chain integration and performance (Shub & Stonebraker 2009). The result of this process implicitly or explicitly fosters increased *cross-corporate understanding* and a *culture of collaboration* throughout the supply chain and in the broader network of firms connected to the firms in the supply chain (Emmett & Crocker 2006, Noorderhaven et al. 2002). The process depends heavily on the *differences* in size and financial and technological *capabilities* among the firms. Therefore, the focal firm in particular needs to stimulate *teamwork*, *goal congruence* and *togetherness* among the firms and people in the supply chain (Fellows 2009). In the end, the process of cultural alignment is influenced to a large extent by *soft factors*, e.g. power balance and trust (Cox et al. 2004, Cox 2004, Ireland 2004). The resulting joint culture is reflected by the notion among the firms in the supply chain to belong to the same '*family*', serve its *joint interest*, and commit to a set of *codes of behaviour*, whether operational, economic or social behaviour. The behaviour of 'family members' promoting communication, commitment and trust is essential in achieving higher degrees of integration, and as a result, higher degrees of *value creation* in the supply chain (Sambasivan & Yen 2010).

Factor of analysis 9: Firms in the supply chain have aligned their respective business cultures to increase commitment and channel behaviour supporting and serving the supply chain as a 'single firm'.

10 Human resource management

As a distinct issue of cross-corporate cultural alignment, firms may align personnel policies to increase *collaborative working* among personnel. This may include forming *joint teams* across the extended business process throughout the supply chain. The joint personnel approach may involve *training and education* of all personnel in the same *supply chain programme* disregarded to which firm in the supply chain they belong. As a result personnel is *interchangeable* between firm which fosters *teambuilding* and *integrated working* in the supply chain. Team building and integrated working are essential to be able to unlock the potential advantage of joint teams and supply chain integration in general. Application of such personnel approaches and team concepts lead to higher levels of *commitment* and *collectivism* on the personal level between the firms in the supply chain (Baiden et al. 2006, Katzenbach & Smith 1993, Kelly & Male 2001). In addition, commitment and collectivism can be promoted by *training* initiatives in the supply chain. This can be facilitated by joint training programmes or requiring employees to attend training programmes offered by other supply chain firms. Relational and joint approaches to training as well as *staffing* have been associated with greater supply chain integration and performance. This is based on the idea of *stabilisation of the workforce* throughout the supply chain and the resulting greater importance of joint HR approaches (Shub & Stonebraker 2009). Supply chain firms may establish joint boards of management, shared meetings and social functions to enhance the *stability of the joint culture* (McAfee et al. 2002).

Factor of analysis 10: Firms in the supply chain have implemented a joint approach to their personnel, including joint staffing and training, fostering integrated working and stability of the workforce throughout the supply chain.

In summary, the four theoretical perspectives on the supply chain result in ten factors of analysis of supply chain integration (Table 12). Both the perspectives and the factors focus on the organisation and coordination, management and governance of the supply chain as it were a 'single firm'. Rather than observing each factor in isolation, the ten factors need to be addressed *collectively* in order to give comprehensive meaning to the concept of supply chain integration, and represent the idea of the supply chain as a 'single firm' usefully.

Table 12: Factors of analysis for supply chain integration

Factors of analysis	Description
Economic governance	
Integration of business activities	The focal firm has integrated business activities internally, and additionally it controls external business activities as well.
Partner sourcing and collaboration strategies	The focal firm has established strategic agreements and long-term collaborative arrangements with key partners in the supply chain.
Production management	
Integration of operations and processes	Firms in the supply chain have connected or integrated the governance of processes and operations to achieve flow, and increase the effectiveness and efficiency of the integrated delivery process.
Planning and logistics control	Firms in the supply chain jointly plan operations and processes, and jointly organise and manage logistics.
Quality management	Firms in the supply chain apply total quality management to processes across corporate boundaries.
Inter-firm governance	
Information exchange	Firms in the supply chain share information and knowledge in a joint information system.
Product development and design	Firms in the supply chain develop and design products jointly and rationally applying product and production modularisation.
Market approach and marketing	Firms in the supply chain are engaged in collaborative market approaches with the aim of delivering integrated products to the end market effectively and efficiently.
Social governance	
Cultural alignment	Firms in the supply chain have aligned their respective business cultures to increase commitment and channel behaviour supporting and serving the supply chain as it were a 'single firm'.
Human resource management	Firms in the supply chain have implemented a joint approach to their personnel, including joint staffing and training, fostering integrated working and stability of the workforce throughout the supply chain.

Isolated approaches to individual factors would lead to incomplete and thus less useful conceptualisations of the supply chain and supply chain integration. For instance, in order to achieve integrated control over the supply chain, the focal firm will either have to exercise direct control by internalising business activities, or gain external control via comprehensive sourcing strategies and collaboration with supply chain firms. For instance, this includes arrangements with regard to joint planning and logistics, and information exchange. In the end, in order to be effective, on an operational level, the integrated workforce in the supply chain will need to support the supply chain integration. Therefore, the respective personnel policies of the separate firms in the supply chain will need to be aligned to achieve a common understanding and culture among the workforce. Essentially, the ten factors of analysis show many mutual *relations*, and thus will be observed as a collective whole, and applied as a coherent model of analysis throughout the remainder of the thesis.

5.2 Identifying four development levels of supply chain integration

Previous approaches to subjects corresponding to supply chain integration have resulted in staged *development* paths or *maturity* models aimed towards 'excellence' in the particular subject (Demeter & Gelei 2003, Li et al. 2005). For instance, on the subject of co-makership, four development or maturity *levels* have previously been defined starting at the level of random selection of suppliers, and continuing to levels of fixed suppliers, and concentration of suppliers, to the ultimate level of 'genuine' co-makership (Broersma 1991). For the subjects of supply chain management and partnership three-staged models have previously been developed assessing applications from low to medium to high levels of advancement (Giunipero & Brand 1996, Lambert et al. 1996). Another model is the maturity model which outlines five maturity levels based on a business process orientation developed previously by the Supply Chain Council (Figure 16).

As an adaptation to the latter model, for this research and the case study analysis in particular, we will use *four levels* of supply chain integration: independent stages, loosely coupled stages, closely connected stages, and ultimately the integrated supply chain. Here we use the term 'stages' to reflect the level of an application, regardless of whether the application includes processes, business activities or firms involved in the supply chain integration. The levels thus broadly reflect an *increasing scale of integration* across an increasing part of the supply chain. However, the level may vary per factor of analysis assessed (Figure 17).

The exact scaling and meaning of the levels per factor will depend per case on the type of firm observed, from clients to suppliers. For a client, for instance, supply chain integration activities include different activities and issues than for a developer, designer, builder or supplier. The levels will be assessed based on the *appearance* of the supply chain integration practice *per case*. The assessment of the different levels of the different cases will therefore be *indicative* and not absolutely comparable. For instance, if cases are on the same level for certain aspects, they will most certainly still differ from each other.

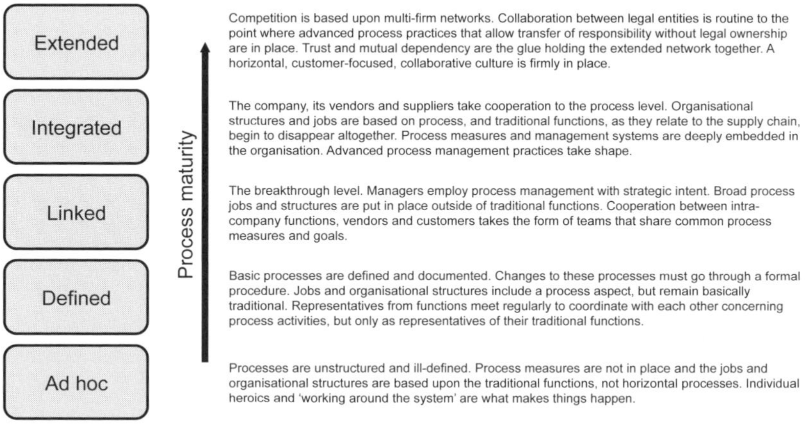

Figure 16: Supply chain management maturity model by the Supply Chain Council (Lockamy & McCormack 2004)

Figure 17: Levels of supply chain integration

5.3 Relating the factors of analysis to the four development levels

As a last development to the analytical framework, the ten factors of analysis are now related with the four development levels, resulting in the *analytical model* that is used for the empirical analysis in the following chapters (Table 13). The four

Table 13: Analytical model for supply chain integration

Development levels	Level 1	Level 2	Level 3	Level 4
	Independent stages	Loosely coupled stages	Closely connected stages	Integrated supply chain
Economic governance				
Integration of business activities	• No activities from other stages in supply chain internalised.	• Few activities internalised from other stages.	• Entire neighbouring stages internalised.	• Multiple neighbouring stages internalised. • Decentralised control.
Partner sourcing and collaboration strategies	• Independent organisations. • Incidental orders. • No incentives. • Specific one-off contracts.	• Regular purchase. • Delivery without ordering. • Limited incentives. • More general contracts.	• Framework agreements. • Annual contracts. • Measured incentives influencing reward. • General long-term contracts.	• Open agreements. • Open cost price. • Risk/reward sharing. • Relationship contracts outlining joint partnering philosophy.
Production management				
Integration of operations and processes	• No coordination of processes between organisations. • No joint investments in production assets. • No cost control. • No time limits.	• Coordination of the interfaces between neighbouring processes. • Few joint assets • Sharing cost data. • Time limits.	• Joint coordination of neighbouring processes. • Input suppliers' know-how in production process. • Multiple joint assets • Joint cost data. • Speeding up processes.	• Integrated processes. • Organisations form an extended enterprise. • Fixed roles. • Assets owned jointly. • Total cost management • Organising for speed.
Planning and logistics control	• Separate planning without coordination. • Logistics organised by suppliers independently. • No inventory control.	• Joint coordination of planning and transport. • Aim for controlling inventory.	• Suppliers have access to and make use of the producer's planning system. • Aim for reducing inventory.	• Open planning system. • JIT logistics managed by suppliers. • Logistics coordination by third parties. • Minimised inventory.
Quality management	• Standard quality. • Separate quality requirements. • Traditional quality inspections by buyer. • Inspection at supplier.	• Specific quality requirements. • Uniform quality levels. • Self-certification. • Free pass to buyer.	• More generic quality guidelines. • Joint quality control with suppliers. • Quality improvement programmes.	• Joint quality philosophy. • Total quality management • Continuous improvement

Inter-firm governance				
Information exchange	• Independent information • No transparency or visibility of information.	• Increased information exchange. • Improved transparency and visibility.	• Facilitated information flow, full exchange. • Same data standards. • High transparency and unlimited visibility.	• Shared ICT, databases, information systems. • All information shared. • Shared knowledge management.
Product development and design	• No involvement of supplier in product development. • Standard products. • No joint R&D.	• Early supplier involvement in new product design. • Discussing design and specifications. • Joint R&D planning.	• Suppliers are fully involved in the design. • Applying design for manufacture to assure manufacturability. • Joint R&D activities.	• Joint product development. • Suppliers' products are standard design parts. • Modular design. • Joint R&D strategies.
Market approach and marketing	• Just marketing of the end product to the end client.	• Suppliers are involved to improve marketing of the end product.	• Suppliers can influence the marketing of the end product including their own products.	• Joint market approach. • Suppliers' products are fully integrated. • Platform strategies.
Social governance				
Cultural alignment	• Different organisational cultures. • Limited trust and short commitment.	• Supporting each other's culture. • Trust building and longer commitment.	• Shared cultural basis. • Trust at high level. • Long-term commitment to each other's operations.	• Shared cultural values. • 'One firm, one mission'. • Implicit total trust. • Commitment to joint long-term success.
Human resource management	• No personnel exchange.	• Limited personnel exchange.	• Extensive personnel exchange. • Inter-firm groups of people working together.	• Joint workforce. • Multidisciplinary teams. • Joint training. • Cross-board managers.

development levels represent a progressive path towards integration per factor of analysis, and towards an integrated supply chain for the factors collectively. The model is based on the general as well as building literature presented throughout the theoretical part of this thesis. Therefore, the model is appropriate for the analysis of the manufacturing as well as the building cases studies in the following chapters.

5.4 Reflection on key issues and contributions of this chapter to the thesis

This chapter represents the closure of the theoretical part of the thesis synthesising the conceptual and theoretical constructs presented in the previous chapters. This has resulted in the deduction of a set of analytical factors which shall function as a basis for the further theoretical development of the concept of supply chain integration for the building industry. In addition to the factors of analysis, levels of supply chain integration have been presented leading to the analytical model to be used for the empirical analysis in the next part of the thesis.

PART II:

EMPIRICAL ANALYSIS

B
232
277

6 Supply chain integration practices in manufacturing

In this chapter, supply chain integration practices as applied in six different sectors of industry are studied: automotive, aerospace, computers, large electronic systems, clothing and grocery. For the analysis, the analytical model from the previous chapter is used to identify and explore the principles of supply chain integration as applied in the six sectors. The exploration includes an observation of the organisational aspects as applied to the delivery process through the supply chain in those sectors. In each sector an advanced company case is investigated as an example. In the course of the research, the manufacturing sector studies and examples in this chapter function as comparisons for the building case studies in the following chapter rather than as case studies in themselves.

6.1 Implications of conceptual investigations across industries

Investigating across industries as done in this research touches conceptual implications such as the comparability and transferability of business and production

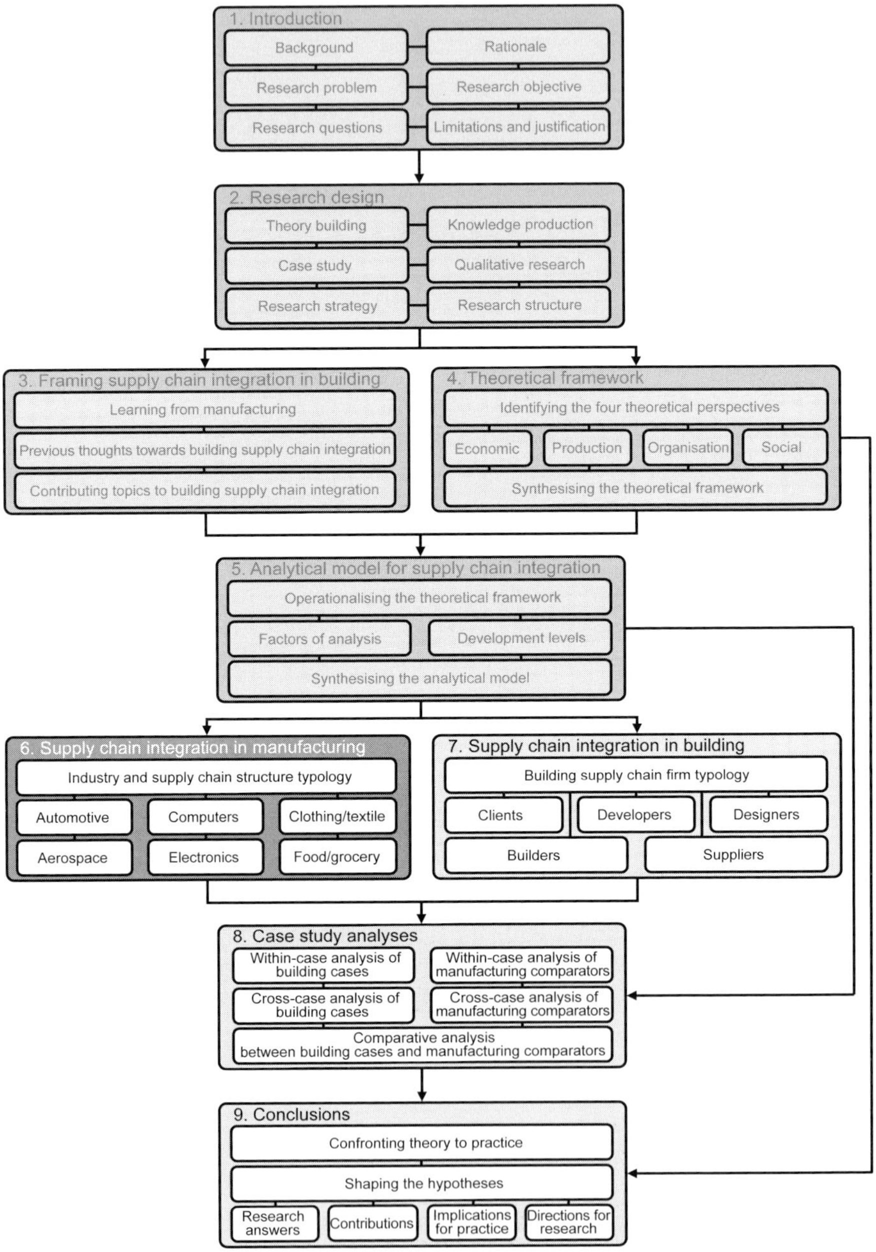
1. Introduction
Background
Rationale
Research problem
Research objective
Research questions
Limitations and justification
2. Research design
Theory building
Knowledge production
Case study
Qualitative research
Research strategy
Research structure
3. Framing supply chain integration in building
Learning from manufacturing
Previous thoughts towards building supply chain integration
Contributing topics to building supply chain integration
4. Theoretical framework
Identifying the four theoretical perspectives
Economic
Production
Organisation
Social
Synthesising the theoretical framework
5. Analytical model for supply chain integration
Operationalising the theoretical framework
Factors of analysis
Development levels
Synthesising the analytical model
6. Supply chain integration in manufacturing
Industry and supply chain structure typology
Automotive
Computers
Clothing/textile
Aerospace
Electronics
Food/grocery
7. Supply chain integration in building
Building supply chain firm typology
Clients
Developers
Designers
Builders
Suppliers
8. Case study analyses
Within-case analysis of building cases
Within-case analysis of manufacturing comparators
Cross-case analysis of building cases
Cross-case analysis of manufacturing comparators
Comparative analysis between building cases and manufacturing comparators
9. Conclusions
Confronting theory to practice
Shaping the hypotheses
Research answers
Contributions
Implications for practice
Directions for research

concepts between different industrial contexts. Although industries are different, applying business and production concepts from one industry to another has been shown to be possible and to provide opportunities for transfer of practices across industries. Some industries are deemed to be more advanced than others in applying sophisticated approaches to production management and supply chain integration. As a consequence, one could say there are opportunities to compare and relate between industries. It has been claimed, for instance, that the building industry could learn from other industries, e.g. the automotive industry for its supply chain approaches, and the defence industry regarding life cycle costing techniques (Garnett & Owen, 1995). However, concepts need to be transferred from the original industry context to the targeted industry context, which may be more difficult for some concepts than for others. Green et al.(2004) warn against being 'over-simplistic' in one's approach, to learn from other industries but not disregard the embedded and contextual nature of managerial concepts. For instance, in the case of a cross-industry investigation between aerospace and the building industry, learning has been observed as 'sharing between business sectors as an essential source of innovation. Comparisons between the aerospace and the building industry sectors are especially useful because they are so different in terms of their institutional context, structure and technological intensity. This helps to explain how managerial practices are mediated by context'.

A recurring issue in debates about the building industry as compared to other industries is that the sector should move away from its ingrained project focus on time and budget, and instead pursue higher levels of speed, innovation, product development and customer focus such as seen in other industries. According to this view, the building industry should also follow a more structured and repetitive approach through projects and supply chains. This implies adopting more integrated and strategic procurement by clients, and applying more integrated supply chain strategies by contractors and suppliers. When it comes to learning from other industries with regard to supply chain integration, the consequences may be that the building industry will need to shift to more integrated and centrally controlled supply chains, accept reduced levels of autonomy among firms in the supply chain, apply product development rather than project development, and deliver repetitive product solutions instead of one-off projects (Voordijk & Vrijhoef, 2003).

6.2 Explanation of the selection of manufacturing industries

6.2.1 Industry typology

Industry characteristics strongly influence the appearance and performance of concepts applied by firms in different industries (Hamblin & Iyer, 1996). The behaviour of firms and the approach to supply chain integration and product delivery in different industries vary considerably depending on the different product and market characteristics (Botter 1980, Wortmann 1992). As a consequence, production channels and the balance between inputs and outputs typically differ per type of manufacturing. Two types of manufacturing are roughly distinguished: process manufacturing (divergent from a small variety of inputs to widely differentiated product outputs, e.g. the food industry), and discrete manufacturers (convergent from large amounts of inputs to specific product outputs, e.g. machine building) (Figure 18). The type of manufacturing shapes the production channel in industries and has a great impact on the production and inventory control of firms, and thus on their approach to the supply chain (Fransoo & Rutten 1994). In process manufacturing, products are often commodities, and firms generally need to be fast and flexible, as does the supply chain. In discrete manufacturing, products are often custom-made, and firms generally need to intensify and stabilise relations in the supply chain.

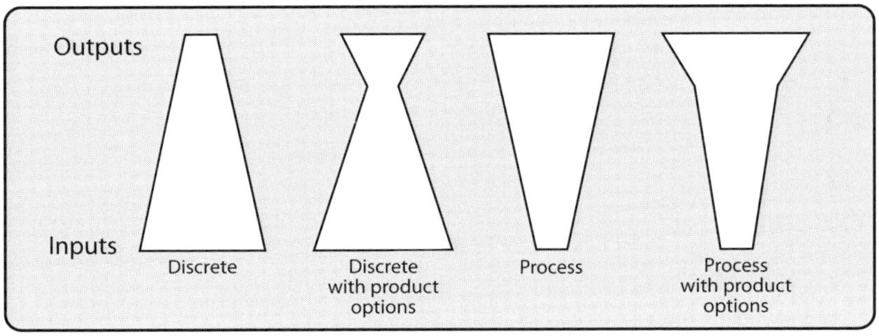

Figure 18: Conceptual differences between process and discrete manufacturing (Fransoo & Rutten 1994)

Recognising that building is particularly related to discrete types of manufacturing, this research predominantly relates to those, i.e. automotive, aerospace, computers and electronics. In addition, two process kinds of manufacturing are observed, i.e. clothing and grocery, in order to distinguish and relate between applications of supply chain integration in both kinds of manufacturing. As a further refinement of this binary industry typology, the supply chain typology below has been added as a second and more specific categorisation for better understanding and discrimination between the different supply chain approaches by the industry sectors and firm examples investigated below.

6.2.2 Supply chain typology

Further to the industry typology, three basic types of supply chain structure can be distinguished, previously identified as Type 1, 2 and 3 supply chain structures by Lin and Shaw (1998). In *Type 1* supply chain structures, the focal firm merely performs the final manufacturing activities such as the assembly of the final product. The suppliers cooperate closely so that less inventory is built up on both sides. Efforts are spent on the cooperation between suppliers and manufacturers and the control of finished product inventory. The manufacturing processes generally rely on capital-intensive equipment and a large number and variety of parts for different products. Such *early differentiation* of product models makes it difficult to satisfy customer-specific demands using the make-to-stock strategy. Therefore, the final products are generated late in the assembly stage. Manufacturing and assembly processes belong to the *convergent assembly* process, where many unique component parts are assembled into a relatively small number of end products (Figure 19).

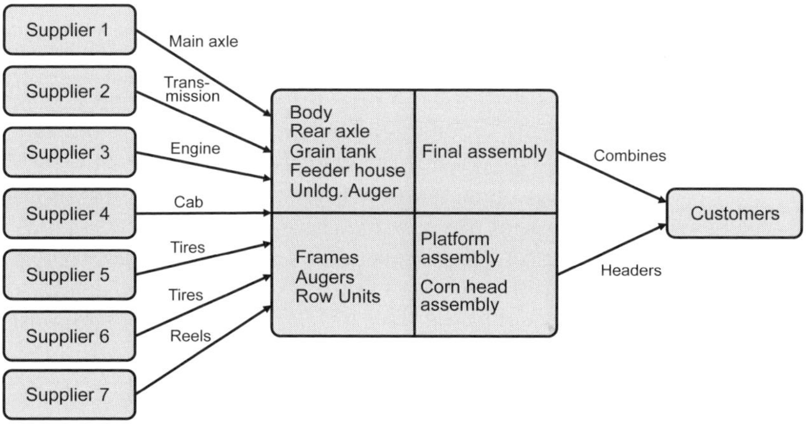

Figure 19: Type 1 supply chain structure: early differentiation, convergent assembly (Lin & Shaw 1998)

In *Type 2* supply chain structures, the focal firm controls and generally owns the final assembly and distribution channel. The assembly process is performed in two stages: first, complex assembly processes for generic models or semi-products are executed, generally at factory sites; then simple assembly processes for customised models are executed at distribution sites. This *delayed differentiation* strategy enables a mass customisation approach. The manufacturing processes following the fabrication and assembly sequence belong to the *divergent assembly* approach, where the component parts are common and can be combined into a large number of different models of end products (Figure 20). The main issue in managing this type of supply chain is to develop strategies to shorten the lead time for customised products.

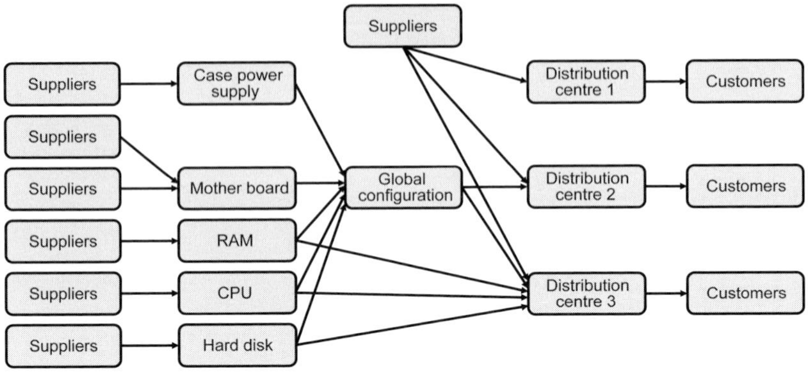

Figure 20: Type 2 supply chain structure: delayed differentiation, divergent assembly (Lin & Shaw 1998)

In *Type 3* supply chain structures, the focal firm may control but not necessarily own the final assembly and distribution channel. Alternatively this may be outsourced to third parties, e.g. retailers. The market environment of this type of supply chain is constantly changing, and *responsiveness* is the key to adapting swiftly to such a dynamic environment. Because the main production process uses a *divergent differentiation* approach, the product models are differentiated at the manufacturing stage (Figure 21). The variety of product models and the quickly changing market shorten the product life cycle. The management of this type of supply chain is concerned with meeting the quickly changing market efficiently by mobilising the supply chain effectively.

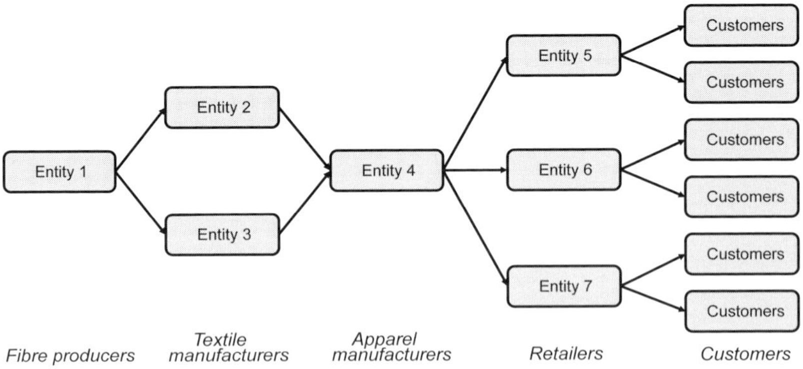

Figure 21: Type 3 supply chain structure: responsiveness, divergent differentiation (Lin & Shaw 1998)

Table 14: Properties of the three types of supply chain. Adapted from (Lin & Shaw 1998)

Attributes	Type 1	Type 2	Type 3
Manufacturing process	Convergent assembly	Divergent assembly	Divergent differentiation
Primary business objectives	Efficiency	Customisation	Responsiveness
Product differentiation	Early	Late	Late
Number of product models	Few	Many	Many
Assembly process	Concentrated at manufacturing stage	Distributed to distribution stage	Concentrated at manufacturing stage
Product life cycle	Years	Months to years	Weeks to months
Main inventory type	End products	Semi-products	Raw materials

The three supply chain types are typical representations of supply chain structures. Just as in the industry typology, the basic difference between the three types of supply chain is the shape of the supply channel from the suppliers up to the customers. In a Type 1 supply chain, more suppliers are involved to design, manufacture, assemble and deliver a specific product to customers. In a Type 2 supply chain, fewer suppliers deliver a greater variety of products, for instance by recombining interchangeable parts, to satisfy different kinds of customers. In a Type 3 supply chain, a relatively small number of suppliers delivers many differentiated products to a large market. The characteristics of the three supply chain types differ not only in structural factors, e.g. the number of tiers and the type of firms, but also in their approach to processes, business objectives and operational issues (Lin & Shaw 1998) (Table 14).

6.2.3 Industries selected for this thesis

Following the above supply chain typology, and recognising the different properties of each supply chain type, six sectors of industry have been selected to explore the applications of supply chain integration in those sectors. First, the automotive and aerospace industries as representatives of Type 1 supply chain structures are explored. Second, the computer and electronics industries as representatives of Type 2 supply chain structures are explored. Third and finally, the clothing and textile industry, and the food and grocery sector as representatives of Type 3 supply chain structures are explored (Table 15).

The six industry sectors and the corresponding example firms will be described below. For each sector, a broad description of the general approach to supply chains and related concepts in the particular industry will be presented. Next, for each sector the corresponding examples will be investigated using the analytical model presented in the previous chapter. This investigation will comprise a descriptive exploration of the application of supply chain integration by the particular firm based on secondary data. An overview of the secondary data can be found in the appendices. The data consists of existing descriptive material including literature of the industry sectors and the firms discussed. The manufacturing cases presented below function as comparative studies rather than case studies, to relate the theory presented in the previous chapters to the building case studies presented in the following chapter. The comparative analysis is found in Chapters 8 and 9, as part of the theory development process of the research, leading to the aim of this thesis, the concept of supply chain integration for the building industry.

Table 15: Manufacturing industries and cases selected for this research

Supply chain structure type	Industries	Examples	Process approach	Manufacturing type
Type 1	Automotive Aerospace	Toyota Airbus	Convergent assembly	Discrete manufacturing
Type 2	Computers Electronics	Dell ASML	Divergent assembly	Discrete manufacturing
Type 3	Clothing and textile Food and grocery	Zara Wal-Mart	Divergent differentiation	Process manufacturing

6.3 Supply chain integration practices applied in the automotive industry

Supply chain integration in the automotive industry traces back to Henry Ford's vertically integrated supply chain system to mass produce cars (Ford 1922, Ford 1926). Ford's vertical integration became a limiting factor in the world of changing customer preferences. Toyota and other Japanese brands perceived suppliers as smaller individual units capable of designing, innovating and adding value to the supply chain (Womack et al. 1990). By disconnecting themselves from the supply chain, car manufacturers aligned themselves to cost and profit management techniques, resulting in greater efficiency and capacity optimisation (Lamming 1993). In the present-day automotive industry, there is a strong reliance on suppliers (Morris et al. 2004). Particularly important is the role of the first-tier suppliers in the supply chain, i.e. the system suppliers of integrated product modules of the car delivered to the manufacturer (Doran 2003).

This influences the coordination of the supply chain as a whole and requires suppliers to be able to 'manage the complexity' associated with modular products, and it can also manage lower-tier suppliers that contribute to the various elements that constitute a module (Doran et al. 2007). Modular coordination and platform strategies are important for the efficiency of the automotive assembly system (Binder &

Clegg 2007). By using a platform approach, a company can develop a set of differentiated products or derivatives (Meyer & Lehnerd 1997, Wheelwright & Clark 1992). Moreover, this approach has enabled manufacturers to synchronise pre- and final assembly activities, share resources and achieve scale effects, and develop activities and resources to increase the efficiency of the assembly system (Frederiksson 2006). As a result of the integrated approach to operations and processes, planning of production and logistics is a precise matter. Suppliers are integrated in the planning systems and logistics procedures to reduce inventory to a minimum, 'pulled' by the assembly line, generally referred to as just-in-time logistics (JIT) (Monden 1994).

Quality in the automotive industry has generally moved to zero defect regimes of 'kaizen' and total quality management (Imai 1986). This requires modules supplied to be zero-defect, too. As part of the modular coordination and total quality management, suppliers are involved in a process of continuous improvement of quality and are responsible for implementing quality improvements in their plants (Fynes et al. 2005a, Fynes et al. 2005b). Knowledge management initiatives have largely been aimed at supplier relations within the automotive industry. Such supplier-focused knowledge management, including its associated processes and procedures, is partly supported by knowledge sharing through face-to-face communication as well as technological knowledge sharing. Supplier involvement in the product development and production process is related to product performance, and effective use of knowledge management tools has an impact on financial performance as well (Lakshman & Parente 2008).

The main design and overall product development of cars is generally still the sole responsibility of the manufacturer, and a crucial factor in competing with other brands. Suppliers are particularly involved in the technological product development of the car and particularly the modules via co-design. Co-design takes place with suppliers via outsourcing of integrated component design and early supplier involvement in product development. Often parallel sourcing takes place with parallel suppliers to develop the same system. The approach includes joint price setting, target costing and profit sharing. The collaboration is organised via information sharing, collaborative design and personnel exchange. Car manufacturers have made efforts to better respond to market conditions via market responsive manufacturing, e.g. responsive sales systems, flexible forecasting, late capacity setting and real-time demand management (Waller 2004).

In many cases, car manufacturers approach the supply base as a family. Suppliers are often brought together on a regular basis for the purposes of knowledge dissemination and group socialisation, e.g. supplier contests for the best performing suppliers. As part of this family approach, many car manufacturers exchange personnel, e.g. guest engineer exchanges between the manufacturer and a supplier or between suppliers (Childerhouse et al. 2003a) (Table 16).

6.3.1 Example of consumer cars: Toyota

Toyota must be seen as the origin of contemporary supply chain integration strategies, as an offspring of the famous Toyota Production System (TPS) (Ohno 1998, Towill 2007), later described in the concept of lean production (Womack et al. 1990, Womack & Jones 1994, Womack & Jones 1996) and eventually translated into a corresponding approach to the supply chain (Lamming 1993, Lamming 1996, MacDuffie & Helper 1997a).

1 Integration of business activities

Internal and external business activities are closely integrated by Toyota. As such, suppliers must be seen as extensions or external business units of Toyota, via long-term partnership, to achieve 'operational excellence', subject to strict cost targets and timing (MacDuffie & Helper 1997b). In fact, all suppliers are located close to the Toyota assembly line to reduce supply risks (Hannon 2008).

2 Partner sourcing and collaboration strategies

Toyota has succeeded in developing a supply chain approach that thrives on close relations with suppliers as an integrated part of their production system. Via mix planning based on forecasts, production stability is achieved. This is translated into sales and operations planning. Sales requests from dealers and resellers 'pull' the production. The pulled production is translated into operations scheduling, parts ordering, supplier management, and logistics orders. Fulfilment of the demand from dealers and resellers closes the loop of the supply chain (Iyer et al. 2009).

3 Integration of operations and processes

The entire production and supply chain functions as an integrated system aimed at balancing variety, variability, velocity and visibility across the entire supply chain to increase value to the customer via the product, and decrease waste in the process (Iyer et al. 2009). Interestingly enough, Toyota has managed to achieve these aims by going beyond large-scale production, and via a lean system of model changeovers, delivering customised products in a highly effective and efficient way (Adler et al. 1999, Lander & Liker 2007, Ohno 1998).

4 Planning and logistics control

Toyota's just-in-time (JIT) planning and logistics are particularly aimed at reduction of production resources, i.e. time, work and inventory, and exposing waste in the process and then removing it. Most of the effort with suppliers is aimed at JIT logistics and setting up support systems such as dedicated transport services and the 'kanban' pull system to enable the assembly line to 'pull' supply (Monden 1994). As part of the JIT system, supplies are organised in small batches with short lead times, and transported with high frequency to the assembly line. Supplies from different suppliers are picked up and transported jointly to the assembly line, also referred to as 'milk rounds' (Lamming 1993).

5 Quality management

Suppliers are involved in quality and learning programmes ('kaizen'), including supplier support and supplier development. The aim is to transfer the TPS philosophy and principles to suppliers, and to create a TPS model in the suppliers' production lines. This includes cost reduction programmes, productivity improvement, inventory reduction, quality improvement, preventive maintenance and housekeeping at the suppliers (Iyer et al. 2009).

6 Information exchange

The knowledge sharing within Toyota's supply network demonstrates Toyota's ability to effectively create and manage a network-structured knowledge-sharing process. Toyota has done this by creating a strong network identity with rules for participation and entry into the network. Most importantly, production knowledge is viewed as the property of the network. Toyota's highly interconnected network approach has established a variety of procedures that facilitate knowledge flows among suppliers (Dyer & Nobeoka 2000).

7 Product development and design

Toyota involves suppliers in new product development with high levels of responsibility for suppliers, based on the position of suppliers in the tiered structure of the supply network, i.e. first-tier system suppliers to lower-tier component suppliers. The first-tier suppliers are especially involved in the development of cars, particularly the fitting in and technological development of the modules of the car, e.g. the cockpit. The development includes strict requirements in terms of quality improvement and cost reductions. A major objective of product planning and design is complexity reduction, for instance, looking for opportunities to use common parts across product models, eliminating options that don't sell well, minimising parts that vary by option and colour, or design accessories that can be installed flexibly after production or at the dealers (Iyer et al. 2009).

8 Market approach and marketing

Toyota's market approach has mainly been dominated by efficiently and faultlessly fulfilling market demands for high-quality cars at relatively low prices. Toyota has also applied 'pacemaking', i.e. developing future technology ready to be installed into cars when the market asks for it, such as hybrid and electric drives. Those technologies in particular have required suppliers, e.g. battery pack suppliers, to be closely involved, Complexity reduction is an objective not only in design but also in marketing, for instance by limiting product offerings for a certain market or region, combining related options into packages, and making high-volume options standard (Iyer et al. 2009).

9 Cultural alignment

Toyota's engagement with suppliers is mainly technologically driven. Choosing a supplier is an intensive and long process with tough conditions to be met.

Table 16: Overview of applications of supply chain integration in the automotive industry, notably Toyota

Factors of analysis	Description of application	Reference to 2nd order themes	Development level
Economic governance			
Integration of business activities	Integration of internal and external business activities. Suppliers as external business units, and close to the assembly line.	Focal firm dominance	4. Integrated supply chain
Partner sourcing and collaboration strategies	Close relations with suppliers. Suppliers organised in a tiered structure. First-tier suppliers responsible for development of integrated modules, and integrating lower-tier suppliers.	Absolute control, strategic collaboration	4. Integrated supply chain
Production management			
Integration of operations and processes	Suppliers are fully integrated in the entire order fulfilment process. The production line of the supplier is fully balanced with the assembly line of the manufacturer.	Operational excellence	4. Integrated supply chain
Planning and logistics control	Supplies of modules to the assembly line are 'pulled' (kanban) and delivered just-in-time (JIT).	Inventory control	4. Integrated supply chain
Quality management	Suppliers are highly responsible for the quality of their own modules and products to fit into the car, including interface coordination with other modules.	Total quality, specialist roles	3. Closely connected stages
Inter-firm governance			
Information exchange	Network-structured knowledge sharing organised by the manufacturer, including all suppliers in an open structure. All knowledge is property of the network.	Teamwork	4. Integrated supply chain
Product development and design	Suppliers are involved early and intensively in product development, particularly first-tier suppliers, particularly the technological development of modules.	Design rationality, specialist roles	3. Closely connected stages
Market approach and marketing	Efficient and faultless order fulfilment. Pacemaking of new technology and gradually pushing this to the market, e.g. hybrid drive.	Technological supremacy, market push	4. Integrated supply chain
Social governance			
Cultural alignment	Supplier development programmes to improve processes and technology. Imposing standards by the manufacturer onto suppliers.	Focal firm dominance, Strategic collaboration	4. Integrated supply chain
Human resource management	Personnel exchange between manufacturer and suppliers. Fostering knowledge exchange and information flow between people in one's own organisation and at suppliers.	Teamwork, specialist roles	4. Integrated supply chain

The process involves extensive verification of whether the supplier has the proper production facilities and R&D capabilities, meets the quality requirements, fits the supply network, contributes to innovation, and is able to decrease cost levels. Toyota engages with suppliers in ongoing learning programmes of continuous quality improvement and problem solving. Toyota demands that suppliers adopt Toyota's standards and puts tremendous pressure on suppliers to perform. Toyota also assists suppliers if needed, particularly with cost cutting programmes (Iyer et al. 2009).

10 Human resource management

As part of the improvement programmes, suppliers' personnel are involved in Toyota's learning and knowledge exchange programmes. Particularly for capability development for the entire conglomerate, Toyota has installed a network of middle-level production specialists from a stable group of firms, who jointly develop and apply new capabilities and technologies (Iyer et al. 2009). Maximum information about those new developments as well as current business is spread to Toyota's and their suppliers' personnel alike (Winfield & Hay 1997). Another key coordination mechanism is the role of guest engineers from suppliers at Toyota, particularly design engineers during product development.

6.4 Supply chain integration practices applied in the aerospace industry

New supply chain concepts applied in the aerospace industry, based on transferring 'best practices' from the automotive to aerospace industry, have proven useful (Mathaisel & Comm, 2000). Production and supply chain concepts in the aerospace industry have been influenced by agile manufacturing, i.e. improving effectiveness and flexibility, as well as lean production, i.e. improving efficiency and flow (Michaels 1999, Phillips 1999). Applications of co-development and co-production between manufacturers and their suppliers have improved the 'value creation' and the competitiveness of supply chains in the aerospace industry (Beelaerts van Blokland et al. 2010).

In the aerospace industry, manufacturers make use of platform strategies. From a production and assembly perspective, the platform makes it possible to share production tools, machines and assembly lines, particularly in the case of complex products, such as aircraft that need integral architectures of major subsystems (airframe, avionics, hydraulics and engines). The aerospace supply chains contain a relatively high number of suppliers. Where product complexity has been reduced this has contributed to reducing the complexity in supply chains and lowering transaction costs. This is achieved through replacing integral architectures with modular ones using a platform approach of interchangeable elements and standardised interfaces.

By applying agile manufacturing in the aerospace industry, the supply chain has turned into a 'common operating environment' (Phillips, 1999; Gunasekaran et al., 2002). Pushing suppliers' flexibility and responsiveness through first-tier level to sub-tier level has been making the aerospace supply chain lean, has reduced variability and production risks, and has centralised commodity management (Michaels, 1999). This has led to applying ever more integrated buyer-supplier relationships in the supply chain (Graham & Ahmed, 2000), to extended enterprise kinds of corporate structures, and to global supply networks (Sehdev et al., 1995; Williams et al., 2002).

Aircraft manufacture is highly dependent on tiers of platform assembly (Williams et al. 2002). Platform assembly has received increasing attention in product development and operations management (Meyer & Lehnerd 1997, Wheelwright & Clark 1992). Components are delivered to fit exactly the structure of the product system and production process of the assembler. The suppliers of components to the aircraft manufacturer are organised in a pyramid of tiers. The first tier supplier integrates all lower tier supplies and develops complete subsystems that fit together with the other subsystems that the assembler eventually assembles into the end product.

Manufacturers aim for minimum lead times from a customer raising a request for a product until it is delivered. Total lead time has to be minimised to enable agility, as demand is highly volatile and thus difficult to forecast. Manufacturers need to respond quickly to exploit market demand. Effective engineering also results in lead time reduction and improvements in cost effectiveness and productivity (Towill 1996c). Particularly in the aerospace industry, quality is an important issue and the subject of continuous effort and control, from the suppliers' plant to the final assembly. Manufacturers have put in place strict rules and checks to assure high levels of product quality throughout the supply chain.

Demand chain management and design for manufacture have led to high levels of structural integration within the supply chain to resolve problems between design and manufacture of complex products and to facilitate direct input of high-tech expertise in the aerospace industry. Current aerospace manufacture is therefore highly dependent on tiers of platform assembly and integral architectures where major subsystems (frame, shell, hydraulics, engines, etc.) are supplied by a relatively high number of closely integrated suppliers. Product technology complexity has increased greatly over the last years, which has evoked efforts to reduce production complexity and increase the level of product modularisation, taking a production platform approach to reduce the complexity of supply chains, e.g. lowering transaction and production costs (Williams et al., 2002). Moreover, manufacturers have tended to involve suppliers' capabilities increasingly, not only for the production of parts and components, but also for the development and engineering of subsystems, such as engines and wings. As a result, the value added has shifted partly from manufacturer to their suppliers, i.e. 'value leverage' (Beelaerts van Blokland 2010) (Table 17).

6.4.1 Example of aircrafts: Airbus

Supply chain integration as applied to aircraft manufacture is further explored by the example of Airbus, and the A380 project in particular.

1 Integration of business activities

Airbus has increased its outsourcing in the A380 program but has retained in-house core technologies such as composite technology and wing design. Airbus's manufacturing, production, and sub-assembly of parts were distributed to around sixteen sites in Europe with final assembly in Toulouse, France and Hamburg, Germany. It had centres for engineering design, sales, and support in North America, as well as and sales and customer support centres in Japan and China.

2 Partner sourcing and collaboration strategies

For the A380, Airbus deployed a vast number of suppliers from various European countries. Airbus looked at supply chain management and outsourcing for better operational efficiencies and found that cost savings from procurement efficiencies could be enormous. In addition, Airbus planned to look for outsourcing to India, China and Brazil, and to produce more aircraft for the local markets in these countries. Airbus pursued new partnering arrangements under the Airbus Power8 restructuring plan. The plan proposed cost-cutting rationalisation measures, closing down or selling specific sites, and rearranging work share allocation. Investment partners were sought for the external production sites in Germany, France and the UK. As part of the Power8 plan, supplier relationships have also changed, because Airbus wanted partners to commit to long-term cost reductions. The company cut out half of its suppliers after 2006. In September 2009, Airbus had 1500 suppliers left, spread across 30 countries. According to the Power8 plan, the number of suppliers was expected to come down to 500 by 2010.

3 Integration of operations and processes

The different components of the A380 were manufactured at different locations – the wings in Wales, the body in Germany, the tail in Spain and the cockpit in France – and then shipped to the Airbus plant in Toulouse for final assembly. This has required strict coordination of the entire operation and the local operations in order to fit the parts together.

4 Planning and logistics control

For the A380 project Airbus invested considerably in logistics. Three special carrier ships were built for Airbus to collect the different parts from Germany, Wales, Spain and France, and eventually ship all parts to the Airbus assembly line in Toulouse.

Table 17: Overview of applications of supply chain integration in the aerospace industry, notably Airbus

Factors of analysis	Description of application	Reference to 2nd order themes	Development level
Economic governance			
Integration of business activities	Keeping core capabilities in house, e.g. wing design. Working with external production locations for major subassemblies. Coordinated via platform strategy.	Focal firm dominance, absolute control	3. Closely connected stages
Partner sourcing and collaboration strategies	Major suppliers involved in risk-sharing partnerships and cost cutting. Supply base has been reduced.	Strategic collaboration	4. Integrated supply chain
Production management			
Integration of operations and processes	Own operations and offsite production locations integrated by central control. Integration of suppliers' processes based on platform strategy.	Absolute control	4. Integrated supply chain
Planning and logistics control	Shorter lead-times by effective engineering of product and subassemblies. Logistics of major subassemblies by large transport equipment.	Inventory control, speed	3. Closely connected stages
Quality management	Strict rules and checks for quality assurance. Approval and surveillance of suppliers' capabilities. Assessments of suppliers' quality management systems by external bodies and use of international quality standards. Central control of quality during engineering.	Total quality, specialist roles, absolute control	4. Integrated supply chain
Inter-firm governance			
Information exchange	Bilateral information control via supply chain information system, ordering online via secured website, collaborative and document exchange design via online engineering environment, sophisticated RFID systems.	Real-time information	4. Integrated supply chain
Product development and design	Suppliers involved in the design of subassemblies for technological input and cost reductions. Manufacturer provides engineering software and centrally controls the engineering.	Design rationality, specialist roles, absolute control	4. Integrated supply chain
Market approach and marketing	No evidence found in the material studied with reference to supply chain integration.	N/A	N/A
Social governance			
Cultural alignment	No evidence found in the material studied with reference to supply chain integration.	N/A	N/A
Human resource management	No evidence found in the material studied with reference to supply chain integration.	N/A	N/A

5 Quality management

One of Airbus's key objectives is to continuously improve its supply chain by winning the suppliers' approval of surveillance of their manufacturing capabilities. To achieve that goal, Airbus has developed a strategy based upon assessments of suppliers' Quality Management Systems (QMS) and their special processes by external bodies, combined with the use of international standards. Once defined, this strategy needs to be deployed throughout the entire supply chain.

6 Information exchange

Airbus suppliers work in parallel, bilaterally with Airbus, with limited lateral communications among themselves. The Airbus supply chain system ensures the right parts; systems and hardware are delivered at the right place on time. The system allows the suppliers and the buyers to exchange the needed information online. The suppliers can display their ability to deliver by filling in the appropriate information on a secured website. Electronic links with suppliers are directed through supplier portals e.g., request for quote/proposal; order placement; technical data interchange, such as technical specifications, key characteristics, engineering drawings; exchanging documents; facilitating virtual collaboration with global partnering suppliers in a 3D design software environment. Airbus has expanded the application of RFID tags for the A380 programs. They have worked together with Boeing to reach a consensus regarding standards for using global RFID technology on commercial airplanes.

7 Product development and design

In order to fit all parts of the A380, the engineering software was provided to all suppliers, and the product development was centrally controlled from Toulouse. Airbus continued to exercise control over all the system and detail engineering down to the component level, including the interface definitions. Major suppliers participated early in the design and development process. Airbus needed to spread development costs, and asked major suppliers in the A380 project to absorb non-recurring costs, thus greatly shifting costs and risks to suppliers. Suppliers were able to be delegated more responsibility for design, development and manufacturing through closer collaboration and integration across supplier networks. Airbus is reported to have established so-called 'risk-sharing partnerships' with many of its major suppliers, covering some 25% of the total non-recurring costs of the programme.

8 Market approach and marketing

No evidence was found in the material studied with reference to supply chain integration.

9 Cultural alignment

No evidence was found in the material studied with reference to supply chain integration.

10 Human resource management

No evidence was found in the material studied with reference to supply chain integration.

6.5 Supply chain integration practices applied in the computer industry

Originally, computer manufacturers' business activities were not integrated outside the assembly of computer systems. In the conventional model, resellers purchased computers from manufacturers and distributed them to the customers. However, led by Dell and Apple, computer manufacturers have begun to integrate the sales and distribution channel, directly selling and retailing end products to the end customer, which has reduced costs and risks of inventories of finished goods, and increased market share and profitability (Magretta 1998a).

Supply contracts for major components in the computer industry usually are tentative contracts drawn for each quarter of the year, in which the buyer makes purchasing commitments to the supplier for that quarter, and approximate times are specified during the quarter when orders will be placed with approximate order quantities. The actual order placing dates and the purchase quantities can be modified as time passes and more information about the demand becomes available (Magretta 1998b, Murty 2000).

Like other sectors of manufacturing, computer manufacturers have increasingly integrated processes and operations in order to achieve operational benefits and cost reduction. This includes suppliers' processes that in some cases are treated as if they were inside the company (Magretta 1998a).

Product and process design methods in the computer industry are dominated by modularity and reduced complexity of product structure in order to enable concurrent and distributed production and increase speed. In the design of the product, these supply chain notions are included, explicitly interlinking the product and the process design in order to optimise the production process and achieve flow production (Krikke et al. 2003). This kind of design for supply chain management includes production and inventory modelling and analysis, supported by sophisticated ICT tools (Garg 1999).

The essential mechanism of flow management is based on increased transparency, increased repetition and organisational learning. These are influenced by the relations between parties, including power, reputation, technological expertise and position of parties in the supply chain in relation to the end customers. Computer manufacturers in particular have taken a leading role here and have gotten more powerful and thus dominant in the information exchange in the supply chain.

Industry-wide modularity and standardisation of components have offered many advantages for producing product variations that have only a limited impact on production and assembly processes, and have led to a reduction of lead times and higher productivity levels (Baldwin & Clark 2000, Watanabc & Anc 2004). In general, modularity has increased computer manufacturers' strategic flexibility and eased product coordination in the supply chain because of interchangeability of parts (Worren et al. 2002).

For high-tech, high-paced industries such as the computer industry, it has become of the utmost importance to bring new technology faster to the marketplace, implying major consequences for the product development, supply, assembly, distribution and sales activities in the supply chain (Hammel et al. 2002) (Table 18).

6.5.1 Example of personal computers: Dell

We will illustrate supply chain integration as applied in the computer industry by looking at the case of a manufacturer of personal computers, Dell Computers.

1 Integration of business activities

Dell's supply chain consists of three stages: the suppliers, the manufacturer, and the customers. Dell's direct contact with customers allows it to properly identify market segments, analyse the requirements and profitability of each segment, and develop more accurate demand forecasts. Global Operations Organization (GOO) is Dell's centre for consolidating its global manufacturing, procurement and supply chain activities. Through GOO, the company aimed to integrate its supply chain and achieve higher efficiency and quality. Overlapping activities would be eliminated, and new manufacturing and distribution models to focus on the requirements of the customers spread across the world would be introduced as a part of GOO. Earlier, all Dell's factories had been managed regionally, and procurement functioned as a separate division.

2 Partner sourcing and collaboration strategies

Dell requires suppliers to be compliant with ISO standards and other quality standards. In order to manage its operations with low inventory levels, Dell collaborates closely with its suppliers. The company's procurement decisions are based on four criteria: quality, cost, delivery and technology. Suppliers are selected on the basis of cost as well as quality, service and flexibility.

3 Integration of operations and processes

Dell saves time on processing orders that other companies would normally absorb in their sales and distribution system. By directly dealing with the customer, Dell gets a clearer indication of market trends. This helps Dell to plan for the future as well as better manage its supply chain. Dell maintains a database to track the purchasing patterns of corporate customers and their budget cycles, in order to forecast demand. It also maintains a database for individual customers in order to cater to their future requirements for computers.

4 Planning and logistics control

Through its forecasting techniques, Dell is able to forecast demand with high accuracy. Thrice per day, the changing demand patterns are communicated to the major suppliers. In all the countries in which Dell operates it has a direct sales force, which is directed by the marketing department located at the headquarters. Dell receives orders via the telephone, Internet, e-mail, etc. Orders are received by business units, which download the orders every 15 minutes. Dell matches supply and demand, and computer configurations are built up from components that are available.

5 Quality management

Dell sets high quality requirements for suppliers. Failing components are analysed by Dell, and the information is fed back to the supplier, who can redesign the component. When suppliers fail to meet Dell's standards structurally, Dell and the supplier will create an action plan to ensure future compliance.

6 Information exchange

Since Dell takes orders directly from its customers, the company reduces the cost of intermediaries that would otherwise add to the total cost. With advancements in technology, the choices available to the consumers have also widened. Customers can use Dell's website to configure their customised computer and place an order for it. Orders go directly to the assembly and suppliers.

7 Product development and design

Using the direct model, Dell provides consumers with tailor-made products, built only after taking the order. In the process, Dell is able to reduce inventory costs and overhead. Through this model, Dell is able to provide its customers with the latest available technologies at a competitive price.

8 Market approach and marketing

Dell pioneered the direct model of selling computers, bypassing the conventional model of selling them through the reseller channel. Dell's strategy is to provide customised, low cost, and quality computers that are delivered on time. Dell successfully implemented this strategy through its efficient manufacturing operations, better supply chain management and direct sales model. Dell believes that by selling computers directly to the consumers, the company will be able to better understand the needs of its customers. Each system is assembled according to the preferences of the customers, at a low price.

Table 18: Overview of applications of supply chain integration in the computer industry, notably Dell

Factors of analysis	Description of application	Reference to 2nd order themes	Development level
Economic governance			
Integration of business activities	Dell in particular applies direct selling and retailing by integration of sales and distribution in the organisation.	Focal firm dominance, market push	4. Integrated supply chain
Partner sourcing and collaboration strategies	Engaging in long-term contracts with suppliers, performance needs to be balanced, incl. service and technological capabilities of suppliers; price is only one factor.	Strategic collaboration, specialist roles	4. Integrated supply chain
Production management			
Integration of operations and processes	Supply, assembly and sales are closely linked. This includes suppliers connected closely to assembly. This also includes the connection between the assembly lines around the globe.	Operational excellence, absolute control	4. Integrated supply chain
Planning and logistics control	Connection with sales and customers enables manufacturers to accurately forecast production, and match demand with supply. Logistics are large-scale and global.	Inventory control, market responsiveness	4. Integrated supply chain
Quality management	Quality standards for suppliers are high. Malfunctioning components require suppliers to redevelop the component. Structural non-compliance is countered with action plans of the manufacturer with the supplier.	Total quality, specialist roles, focal firm dominance	4. Integrated supply chain
Inter-firm governance			
Information exchange	Direct information is passed from the customer ordering a computer, to assembly and suppliers. Manufacturer is dominant in the information exchange.	Real-time information, focal firm dominance	4. Integrated supply chain
Product development and design	Products are assembled to order based on customer demand. Direct connection with the customer enables development of new products, involving suppliers.	Design rationality, market responsiveness	3. Closely connected stages
Market approach and marketing	Direct sales to customers and delivering customised computers according to customers' wishes, fast and for a low price.	Speed, market responsiveness	4. Integrated supply chain
Social governance			
Cultural alignment	Manufacturer put in place supplier development programmes, including supplier conferences, suppliers subscribing to manufacturer codes, suppliers doing self-audits	Focal firm dominance, strategic cooperation	3. Closely connected stages
Human resource management	No evidence found in the material studied with reference to supply chain integration.	N/A	N/A

9 Cultural alignment

Dell provides a platform to its suppliers, for instance, through its annual supplier conference, publications, and training in a variety of areas such as environmental practices. As part of every new supplier contracting process, Dell requires suppliers to agree to Dell's requirements and principles. In order to embed socially responsible behaviour into business activities, Dell includes a review of requirements and principles in quarterly business reviews that are held with key suppliers. Dell asks key suppliers to conduct a self-audit to review with Dell management on an annual basis, using a standard scorecard format.

10 Human resource management

No evidence was found in the material studied with reference to supply chain integration.

6.6 Supply chain integration practices applied in the electronics industry

Supply chain management has been a key issue in the production of electronics. The increasing speed of markets and fast obsolescence of digital parts and equipment, such as semiconductors, set high requirements for supply chains. Fast and effective channel management has become a key business strategy, not only to increase speed of production itself, but also to be able to be able to develop new products and adapt existing products in the most efficient way. Flow management of material, information and resources among parties along the supply chain must be aimed at maximising responsiveness and minimising variability and uncertainty (Berry et al. 1994).

The electronics industry has been widely transferring production operations to countries with lower cost levels for reasons of resource efficiency (Jiang & Hansen 2003). In the electronics industry, collaborative arrangements between parties in the supply chain are of essential importance for technological and competitive reasons. Because the pace and the technological standards in the industry are so high, and the market is growing increasingly global, no single firm can operate on its own. Parties in the electronics supply chain are highly interdependent. Collaborative networks of parties must be committed yet flexible enough to operate and interact optimally as well as being able to adapt to new technologies and new competitive situations in the industry and the global marketplace. Collaborative networking should therefore be structured in an adaptive way for 'vertical cooperation' through multiple stages of the industrial channel through the supply chain (Fujimoto 2003). In order to achieve this, new models of industrial organisation have been introduced to the industry, e.g. modular production networks, where manufacturers focus increasingly on the 'creation, penetration and defence of markets for end products and services, while sourcing their operational capacity out to globally operating key suppliers' (Sturgeon 2002).

In the supply chain structure, from a focal firm's point of view, this implies achieving certain levels of vertical disintegration and instead establishing 'close relationships with suppliers governed within a quasi-vertically integrated supply chain' (Webster et al. 1997), so as to respond to changes in the marketplace more effectively and efficiently. Strategic outsourcing alliances with suppliers and contract manufacturing have increased the agility of electronics supply chains (Mason et al. 2002). Various factors, such as the technological character of the parts supplied, cost levels, etc., determine the proximity of the suppliers and the decision to install a regional supply network (industrial district) (Rama et al. 2003) or an international supply network (Chen 2003). Integrated order and inventory management and partnering sourcing are management principles that must be in place to achieve the goals of flow management. In this sense, flow management is related to notions of lean supply and value stream management to achieve a production process with maximised 'value creation' and minimised 'waste' (Hines et al. 1999).

In high-paced sectors of the electronics industry, such as semiconductors, the reduction of lead-times and inventory levels is of essential importance for reasons of cost levels and product obsolescence. Various techniques are applied to achieve these goals, including work-in-progress balance control and bottleneck scheduling (Lee & Kim 2002). Again, all these techniques are aimed at the optimisation of the production process throughout the supply chain. In the electronics supply chain it is essential that information is shared and that the information is reliable, particularly planning information. It is essential that parties along the supply chain can anticipate changing situations, such as fluctuations in the deliveries, particularly delays (Ren et al. 2003). Therefore, an increasing array of business process improvement programmes has been implemented across the industry, including just-in-time management, logistics integration, vendor management and time-based management (Mould & King 1995, Towill 1996a, Towill 1996b, Towill 1996c).

Besides speed and flexibility, the quality of parts and supplies has become of increasing importance because of the ever increasing complexity of technology. Therefore, firms have applied total quality management, particularly with suppliers (Levy et al. 1995).

Manufacturers have generally become large and powerful players in the electronics industry and thus of great importance for information exchange. On the other hand, contractual relations with other parties in the supply chain have been a major influence on information flows in the supply chain. Sophisticated information systems and strict information management have been essential factors in achieving information exchange.

Modularity of products has increased the overall performance of firms in the electronics industry. The establishment of industry standards has played a major role in the widespread adoption of modularity principles (Baldwin & Clark 2000, Gadde & Jellbo 2002). Product modularity combined with platform strategies has enabled products to be differentiated to a high degree, meeting varied customer requirements, i.e. multi-objective optimisation within a reconfigurable manufacturing system (Rai & Allada 2003, Yigit & Allahverdi 2003).

In the electronics industry, speed has become a major driving force of growing importance for global competition. It is an extra complicating dimension for international firms competing on a global scale as competition has changed, for instance, because of large, Chinese firms joining the global electronics market. Therefore, business process re-engineering is a continuous process for electronics firms to stay competitive and survive in a changing global environment (Berry et al. 1999).

Inter-organisational relations within the electronics supply chain have developed considerably. The scope and duration of inter-organisational relations have widened, and therefore issues such as competitive relations, power balance, reciprocity, mutuality and opportunism have become more important (Fynes & Ainamo 1998, Kim et al. 1999) (Table 19).

6.6.1 Example of semiconductor machines: ASML

1 Integration of business activities

ASML has traditionally produced only some 10% of its end product internally. This has been so from the company's beginning and is a difference from its main competitors, Canon and Nikon, who are both assumed to be highly vertically integrated. The bulk of production is outsourced to, broadly speaking, three types of suppliers: producers of optical equipment (lasers, lenses), mechanical equipment, and electronic equipment. The company has some sixty key, first-tier suppliers, but its entire supply network is easily more than ten times that. With the bulk of its key suppliers, ASML accounts for less than 25% of these companies' revenue. Nevertheless, there are some companies or business units of larger companies where the dependency on ASML orders is considerably higher.

2 Partner sourcing and collaboration strategies

The managerial attitude of ASML has been relatively opportunistic, including in its behaviour towards its supplier base. Nevertheless, there has been a considerable emphasis on developing long-term relationships with key suppliers. As a result, several if not most of ASML's key suppliers today are the same ones that the company started off with years ago.

3 Integration of operations and processes

No evidence was found in the material studied with reference to supply chain integration.

4 Planning and logistics control

Based on the detailed forecast, a plan is generated in the ERP (Enterprise Resource Planning) system, and using MRP (Manufacturing Resource Planning) logic, the forecast is translated to component requirements. Suppliers will receive orders as future requirements exceed total pipeline availability. If, however, the forecast is not correct, it might be possible that components requirements drop. The MRP logic at ASML can then reschedule the due dates of the orders at the supplier to a date upon which the items are currently required.

5 Quality management

Quality requirements are very high, including for parts of the systems that ASML produces. For instance, lens quality plays a key role in determining the system's resolution capabilities. Together with ASML's partner Carl Zeiss, ASML is continually striving to improve the quality of its lenses.

6 Information exchange

The supply chain departments are responsible for creating supply possibilities by following the determined business scenario as closely as possible. First, the information gathering process is described. Second, the processes after the decisions are discussed. To be able to determine the state of the supply chain, it is necessary that information can be collected instantly. ASML and suppliers will together keep the information updated, preferably by standardised interfaces. The information is used to determine opportunities to supply goods to ASML customers. A project is defined to write the process and procedures for storing and sharing information. In this project, the requirements for the data are set as well as the amount of data which is shared, how the data is shared, and at what frequency the data is shared.

7 Product development and design

ASML develops platform-based product families for its whole product range. It gives two main reasons for following a family approach. First, a stable platform makes it easier to come up with newer modules and to ramp up volume. Second, from an engineering point of view, it is unaffordable to design a new machine from scratch every time a change in a local part of the machine is needed. Besides efficiency in the development process, benefits include shorter time-to-market and ramp-up times, advantages for servicing and maintaining the machines, and improved learning curves.

8 Market approach and marketing

ASML is the market leader in semiconductor machines. These are complex machines in a low volume market where lead times for components and assembly are much longer than the customer is prepared to wait. ASML makes detailed forecasts over a horizon longer than a year ahead.

Table 19: Overview of applications of supply chain integration in the electronics industry, notably ASML

Factors of analysis	Description of application	Reference to 2nd order themes	Development level
Economic governance			
Integration of business activities	Fast and effective channel management has become a key business strategy, not only to increase speed of production. Key suppliers are heavily involved.	Focal firm dominance, speed	3. Closely connected stages
Partner sourcing and collaboration strategies	Sourcing strategy is aimed at developing long-term relationships with key suppliers	Strategic collaboration, specialist roles	3. Closely connected stages
Production management			
Integration of operations and processes	No evidence found in the material studied with reference to supply chain integration.	N/A	N/A
Planning and logistics control	Planning is aimed at reduced lead times and time to market. Resource planning systems support forecasting, and orders and scheduling towards suppliers.	Inventory control, speed	3. Closely connected stages
Quality management	Total quality management and continuous improvement with suppliers to improve parts.	Total quality, specialist roles	4. Integrated supply chain
Inter-firm governance			
Information exchange	Information is mainly exchanged to increase market responsiveness.	Market responsiveness	2. Loosely coupled stages
Product development and design	Product development based on platform strategies and reconfiguring existing products to increase efficiency, control investments, and shorten time to market.	Design rationality, speed, technological supremacy	4. Integrated supply chain
Market approach and marketing	Speed and good forecasting are essential aims while lead times of complex systems are often long	Speed	3. Closely connected stages
Social governance			
Cultural alignment	Efforts are aimed at building trust and transparency and sustaining long-term relations with suppliers, although the market and as a result the orders are highly volatile	Strategic cooperation, commercial focus	3. Closely connected stages
Human resource management	No evidence found in the material studied with reference to supply chain integration.	N/A	N/A

9 Cultural alignment

Mutual trust and transparency play an important part in this management policy. This trust has been severely tested in the past due to the steep ramp-ups and ramp-downs that characterise the semiconductor industry, where year-to-year changes of over 40% are not uncommon. Suppliers are surprised by sudden decreases in ASML's orders when the business cycle turns downward and face considerable supply chain difficulties as a result. As a result, some suppliers are reluctant to trust a steep increase in orders as the business cycle turns up again. Despite ASML's emphasis on long-term relationships with suppliers, it is with these sceptical suppliers that the company's level of business decreases over time. Also, the three-year business cycle and long-term growth rate of around 20% fit with the incoming customer order rate in the simulation model for this study.

10 Human resource management

No evidence was found in the material studied with reference to supply chain integration.

6.7 Supply chain integration practices applied in the clothing and textile industry

Increasing internationalisation and integration have had a major impact on buyer-supplier relationships and inter-organisational business relationships in the clothing and textile industry (Crane 1997, Da Villa & Panizzolo 1996). Supply chain management in the clothing and textile industry has evolved from transaction governance toward more hybrid market-hierarchy arrangements to economise on transactional costs, increase flexibility and dynamic efficiency, and thereby improve the competitive positions of firms in the sector (Mariotti & Cainarca 1986). The development has led to the emergence of inter-organisational forms such as strategic groups (McNamee & McHugh 1989), enterprise clusters (Visser 1999), and industrial districts (Pietrobelli & Barrera 2002). However, vertical integration is still applied by certain fashion houses. Increased internationalisation has had major implications for the organisation and governance of clothing and textile supply chains, and on purchasing and relations with suppliers (Meijboom 1999).

In search of cuts in lead times and cost reduction in production and distribution, various well-known strategic principles have been applied, such as quick response (QR) or accurate response (AR) (Chandra & Kamur 2000). These strategies have become a prerequisite for doing business in the contemporary clothing and textile industry (McMichael et al. 2000, Meijboom 1999). Lead times of supply and throughput have been compressed to a minimum in order to achieve QR (Forza & Vinelli 2000). Partnership relations with distributors and strong management must be aimed at lead time compression.

The emergence of global commodity chains and international operations networks, and the shift from traditional assembly of imported inputs to a more regionally integrated and higher value-added manufacturing, have encouraged exporters to establish close linkages with many different manufacturers (Gereffi 1999). Different types of manufacturers use different networks and source in different parts of the world. Retailers and exporters tend to rely more and more on regional full-package sourcing networks, in which they buy ready-made apparel primarily from manufacturers (Bair & Gereffi 2001). However, manufacturers have tended to develop multi-layered global sourcing networks where low-wage assembly can be done in other parts of the world. However, keeping a portion of the manufacturing onshore at an agile, quick response base has been suggested to be more cost effective, and increases competitiveness (Warburton & Stratton 2002). In a search for lower costs and improved competitiveness, some manufacturers have evolved into 'almost virtual organisations with outsourcing being its prime activity, e.g. Benetton and Tommy Hilfiger' (Sparks & Wagner 2003).

Inventory management and inventory decision-making are of strategic importance in clothing and textile supply chains (Chandra & Kumar 2001, Raman 1999). Inventory management and decision models anticipate variations in demand due to consumer behaviour by incorporating techniques to fulfil these diverse demands and to translate them into the context of production and distribution. This requires a 'pull system' to support a QR strategy with higher levels of inventories based on product demand patterns, mandated to avoid stock-outs, and thus resulting in higher costs (Chandra & Kumar 2001).

The successful adoption of quick response (QR) practices and technologies in the clothing and textile industry depend on various factors: high quality, flexibility, strict adherence to agreed-upon goals throughout the supply chain, the adoption and implementation of information transference technologies and operating procedures, developing trust between firms in the supply chain and fostering working relationships or collaborations (Perry & Sohal 2000). These factors have also been observed to be important for improved quality and the performance of the clothing and textile industry in general (Perry & Sohal 2001, Sohal et al. 1998).

Efficient inventory management has been made possible to a large extent by the introduction of microelectronics such as RFID in the production and distribution of clothes and garments (Hoffman 1985). The introduction of inter-organisational information systems across organisational boundaries has had a great impact on the structure and management of supply chains, cooperative relationships and competitive positions. These systems have supported the integration of supply chains towards 'electronic hierarchies', and shifting competition more and more between separate supply chains rather than between individual firms (Holland 1995). As a result, information technology and e-commerce play an increasingly important role (Loebbecke et al. 1996, Tatsiopoulos et al. 2002, Yen 2002) (Table 20).

6.7.1 Example of fashion: Zara

1 Integration of business activities

Zara operates using a vertical supply chain, which currently is a unique strategy in the fashion industry. Vertically integrated business undertakes a variety of activities from designing, sourcing and manufacturing, to distribution to retail stores around the world. A company that operates according to a vertically integrated strategy has total control of its various business activities and its business management. The key to Zara's success has been this vertically integrated structure where design, production, distribution, and retailing were integrated. Zara's vertically integrated supply chain system enabled the company to place the latest designs in any store across the world within a period of two to three weeks. The company produced garments according to the latest trends in a limited quantity. Zara introduces 12,000 designs every year, with new designs appearing in the stores globally twice a week. The design, production and distribution of a new series of product can take place in just 15 days. This speed decreases the amount of work in progress and the working capital, and thus increases return on investment and makes Zara's profits higher than those of competitors. The high speed allows Zara to work with extra capacity, in fact a slight overcapacity, in order to maintain its responsiveness and speed at a constantly high level.

2 Partner sourcing and collaboration strategies

The balance between fabric supply and manufacturing of the product in itself contributes to the company's success. The major part is done in house, but not all the company's in-house resources are used. A large percentage of the company's fabric supplies are used by other brands. More than half of the products are fabricated in house. The remainder is outsourced to another firm who is largely dependent on Zara as a client. This enables Zara to optimally coordinate the supply chain, particularly as compared to competitors, who have often outsourced virtually all fabrication.

3 Integration of operations and processes

The company's success is due to its total control of every aspect of the business, from designing to production to distribution. By having total control of the entire process, the company can monitor and quickly react to rapidly changing fashion trends and customer taste. All of the functions of the business continuously work together to produce new collections and designs which are updated and completed on a weekly basis. Shop managers report to designers on a daily basis about what has and has not sold. This report is used to determine if a product is to be kept or altered, and whether new lines are to be created. This happens within a few days. The designers rely primarily on product sales, feedback and comments from customers. Stores order their stock from an offering they receive twice a week from the commercial manager who then passes the orders on to the logistics centre for handling. Stores are ranked according to sales and forecast accuracy; this rank will determine the level of priority for a store order. If a product is not selling in a particular store, the company halts the manufacturing. This prevents stock piles of non-selling products. Instead of maximised production, Zara chooses to maintain considerable overcapacity in its factories. Production is done in small batches based on strict ordering and order taking by the shops. And even in the shops, part of the space is deliberately kept empty.

4 Planning and logistics control

The relentless introduction of new products in small quantities reduces the costs of running out of any particular item. Zara makes a virtue of stock-outs. Zara has an informal policy of moving unsold items after two or three weeks. This can be an expensive practice for a typical store, but since Zara stores receive small shipments and carry little inventory. All of Zara's products are transported from the company's main central site in Spain. Most of the products are shipped from the central depot Zara Logistica. Stock is not held for long periods but is sent out to the Zara stores twice a week. For international deliveries, the stock is delivered to the border of Spain, and the logistics provider in charge for that country takes over the distribution to the stores. Zara is in control of all logistics and the coordination of most of its shops, and thus of all inventory and replenishment. The sales, production and supply chain act as one, so no unnecessary inventory is created. Transport and deliveries follow a strict schedule of twice per week, with deliveries at exact times. If a shop is late with ordering, it will have to wait for the next scheduled delivery. All products are labelled and ready to be put in the shops. This schedule is maintained by planning overcapacity into the logistics system, so delays and unforeseen situations can be buffered in the system, and deliveries are guaranteed.

5 Quality management

Zara is driven by introduction of new designs and releasing new products in a short time. The strategy is to produce and release products in a limited number per store; a store may only receive ten of a new product. This strategy closely emulates a 'make to order environment'. The strategy of releasing a design in limited quantity or exclusively builds up customers' anticipation of the next product or design to be released, making the next product highly anticipated by customers. Zara does not focus on advertising their product, because Zara does not focus on building brand image; their target is production and the customer's anticipation of their product. Instead of focusing their strategy on product advertisement, the company focuses on product design and quality.

6 Information exchange

Zara's 'fast fashion' system depends on a constant exchange of information throughout every part of Zara's supply chain, from customers to store managers, from store managers to market specialists and designers, from designers to production staff, from buyers to subcontractors, from warehouse managers to distributors, and so on. Most companies insert layers of bureaucracy that can bog down communication between departments. But Zara's organisation, operational procedures, performance measures, and even its office layouts are all designed to make information transfer easy. Zara is careful about the way it deploys the latest information technology tools to facilitate these informal exchanges. Customised handheld computers support the connection between the retail stores and the main office. These PDAs augment regular (often weekly) phone conversations between the store managers and the market specialists assigned to them. Through the PDAs and telephone conversations, stores transmit all kinds of information to the main office, such as hard data on orders and sales trends and soft data such as customer reactions and the 'buzz' around a new style. Once the team selects a prototype for production, the designers refine colours and textures on a computer-aided design system. If the item is to be made in one of Zara's factories, they transmit the specifications directly to the relevant cutting machines and other systems in that factory. Bar codes track the cut pieces as they are converted into garments through the various steps involved in production (including sewing operations usually done by subcontractors), distribution, and delivery to the stores, where the communication cycle began. The constant flow of updated data about client buying behaviour, trends and sales mitigates the bullwhip effect. In this way, Zara avoids costly overproduction and the subsequent sales and discounting.

7 Product development and design

Zara has centralised design at its design centre consisting of three halls: one for women's, one for men's, and one for children's clothing lines. Unlike most companies, Zara makes a point of running three parallel, but operationally distinct, product families. Accordingly, separate design, sales, procurement and production-planning staffs are dedicated to each clothing line. The physical and organisational proximity of the three groups increases both the speed and the quality of the design process. Designers can quickly and informally check initial sketches with colleagues. Market specialists, who are in constant touch with store managers, and many of whom have been store managers themselves, provide quick feed back about the look of the new designs (style, colour, fabric, and so on) and suggest possible market price points. Procurement and production planners make preliminary, but crucial, estimates of manufacturing costs and available capacity. The cross-functional teams can examine prototypes in the hall, choose a design, and commit resources for its production and introduction in a few hours if necessary. Ideas for new designs or for modifications to be made to existing designs mainly come from Zara's stores. Instead of projecting sales for a certain colour, fabric, or style and launching such products, Zara reacts swiftly to emerging trends in the fashion industry. The company ensures that its stores are stocked with the products that the customers want at that time.

8 Market approach and marketing

In Zara stores, customers can always find new products, but in limited supply. There is a sense of exclusivity, since only a few items are on display even though

Table 20: Overview of applications of supply chain integration in the clothing and textile industry, notably Zara

Factors of analysis	Description of application	Reference to 2nd order themes	Development level
Economic governance			
Integration of business activities	Generally supply chains are less integrated to be flexible. However, some fashion houses apply vertical integration, internalising all business functions in house.	Focal firm dominance, absolute control	4. Integrated supply chain
Partner sourcing and collaboration strategies	Sourcing strategies are aimed at cost efficient procurement of fabric. Often purchasing is organised globally in order to reduce costs.	Commercial focus	3. Closely connected stages
Production management			
Integration of operations and processes	Processes are highly integrated because of responsiveness, speed, inventory and the need for information about customers' preferences and new trends.	Operational excellence, market responsiveness, speed	4. Integrated supply chain
Planning and logistics control	All efforts are aimed at speed and strict control of inventory, preventing stock-outs or oversupply.	Inventory control, speed	4. Integrated supply chain
Quality management	Quality is standard routine and set as a straightforward target for suppliers to meet.	Focal form dominance	2. Loosely coupled stages
Inter-firm governance			
Information exchange	Information exchange is real-time to support logistics and materials supply, and to capture information about new trends and sales from the shops. Supported by technology, such as barcodes, microelectronics and e-commerce.	Real-time information, inventory control, market responsiveness	4. Integrated supply chain
Product development and design	Multi-functional design teams, including market experts and salespeople with knowledge of new trends, international fashion, etc. Postponement of the actual production assures clothes being produced meet the latest trends.	Teamwork, design rationality, responsiveness	4. Integrated supply chain
Market approach and marketing	Low volumes of products to maintain a level of scarcity and exclusivity, influencing buying behaviour of customers. Global regions are approached differently, products adjusted to meet regional preferences. Releasing new products constantly.	Commercial focus, speed, absolute control	4. Integrated supply chain
Social governance			
Cultural alignment	No evidence found in the material studied with reference to supply chain integration.	N/A	N/A
Human resource management	No evidence found in the material studied with reference to supply chain integration.	N/A	N/A

stores are spacious. Such a retail concept depends on the regular creation and rapid replenishment of small batches of new goods. Zara's designers create approximately 40,000 new designs annually, from which 10,000 are selected for production. Some of them resemble the latest couture creations. But Zara often beats the high-fashion houses to the market and offers almost the same products, made with less expensive fabric, at much lower prices. Since most garments come in five to six colours and five to seven sizes, Zara's system has to deal with something in the realm of 300,000 new stock-keeping units (SKUs), on average every year. Each store has different customers or segments. Each country and each region have different values. Customer segments have different preferences or values in terms of their product choice. Because products are released in a limited way, customers regularly visit a store to see if new stock has arrived. Some products in a region may not be as high selling as in another region. In the Asian market, some designs are kept and maintained, while in most European countries where fashion is in constant flux, designs must be constantly new, therefore requiring the constant release of new products. Customers in other regions tend to embrace whatever is the latest design in Europe, it being considered the fashion capital of the world. Success in the European market proves that a product can be marketed elsewhere, such as in Asia.

9 Cultural alignment

No evidence was found in the material studied with reference to supply chain integration.

10 Human resource management

No evidence was found in the material studied with reference to supply chain integration.

6.8 Supply chain integration practices applied in the food and grocery sector

Supply chain management is the key to the successful control of logistical activities in the food and grocery sector (Christopher 1992), particularly for perishable food products (Voordijk et al. 1999). On the one hand, the 'stability' of the supply system, whether in transportation, storage or packaging is crucial for the quality of food such as vegetables (Ballou 1998, Minegishi & Thiel 2000). On the other hand, market developments force the food as well as the grocery supply chain to respond quickly to rapidly changing market needs.

Supply chain management of food and groceries has become more complex because of the increasing diversity of consumer preferences, globalisation of markets and technological innovations. The market for these products has transformed radically from a seller's to a buyer's market (King & Phumpiu 1996). This development demands supply chains to respond in a quick and highly frequent manner to changing market needs. Products have to be delivered within a manageable and reliable lead time to a growing diversity of outlets. In order to provide for such logistical demands as responsiveness, controllable and reliable lead times, and a high frequency of delivery to a great diversity of outlets, the application of several logistics, information and production management concepts has improved the management of food and grocery supply chains. The growing complexity and volatility of the business has increased the uncertainty and has driven food and grocery supply chain towards vertical integration.

Higher levels of information among parties along the supply chain could relieve the uncertainty and thereby become a substitute for vertical integration (Loader 1997). The nature of logistics and supply chain management in the food and grocery sector has changed as a result of new information technology enabling firms to transform their dealings with their customers and suppliers. Interrelated concepts such as Quick Response (QR), Efficient Customer Response (ECR), Cross-docking and Collaborative Planning, Forecasting and Replenishment (CPFR) have been increasingly utilised by the food and grocery sector to promote supply chain performance. These customer-focused strategies encourage full cooperation, strategic alignment and strong relationships between retailers, suppliers and distributors, working towards mutual goals of reduced cost and greater efficiency (Sparks & Wagner 2003).

ECR is an important concept in food and grocery supply chains (Van der Vorst 2000). It is a partnership strategy in which the retailer and supplier work together to respond more quickly to consumer needs by sharing sales information in order to jointly forecast future demand. The principles of QR were translated to the grocery supply chain, which led to ECR strategies in which distributors and suppliers work closely together to offer better value for the consumer. By jointly focusing on the total grocery supply system rather than the individual components, they reduce total system costs, inventories and physical assets while improving the consumer's choice. Postponement of operations, e.g. assembly, has been identified as a component of ECR enabling better response to demand changes, and consequently improve the performance of food supply chains (Van Hoek 1997, Van Hoek 1999). In addition, product range management can contribute to ECR where different markets, e.g. regional differences, necessitate product variety, and the ability to respond quickly and efficiently across the full product range, for instance by separating operations (Holmström 1997).

Cross-docking and joint distribution pool manufacturers' distribution facilities and shorten the lead time in order to reach more customers in time. By cross-docking, products from various factories/auctions are received in one distribution centre (DC) and shipped the same day to the DCs of the retailers. Because of the growing product diversity and the decreasing average stock per product at the supermarkets and in the retailers' DCs, suppliers are compelled to deliver smaller product quantities more frequently. In order to increase transport efficiency, products are collected from various factories and auctions, and transported to a central cross-docking location of a logistics service provider or a group of factories. From this location, the retailers' DCs are supplied.

The implementation of QR and ECR strategies in the food and grocery sector has implied the implementation of robust inter-organisational information and logistics management systems, including CPFR. Application of CPFR implies collaborative planning and mutually agreed upon forecasts based on the best data available to the parties involved. Results have included improved trading relationships and streamlined supply chain processes (Sparks & Wagner 2003). In some cases, the responsibility for the inventory management and the distribution is shifted entirely to the supplier through a vendor managed inventory (VMI) system (Tyan & Wee 2003). These developments increasingly require on-line logistics service systems (Rae-Smith & Ellinger 2002), e.g. e-grocery (Tyan & Wee 2003), e-marketplaces (Eng 2003) and electronic data interchange (Hill & Scudder 2002) (Table 21).

6.8.1 Example of groceries: Wal-Mart

1 Integration of business activities

Rather than internalising business activities, Wal-mart prefers to establish partnerships, and concentrates itself on logistics and sales.

2 Partner sourcing and collaboration strategies

Over the past decade, Wal-Mart's practice of inviting its major suppliers to jointly develop powerful supply chain partnerships has gained attention. These partnerships are designed to increase product flow efficiency and, consequently, Wal-Mart's profitability. Many companies have stepped up to the challenge, starting with the well-known Wal-Mart/Procter & Gamble alliance, which incorporated vendor-managed inventory, category management, and other intercompany innovations. Procter & Gamble (P&G) even fielded a dedicated account team. In a very creative approach, the team members represented key P&G functions: sales/marketing, distribution/supply chain management, IT and finance. In the eyes of one P&G vice president who was pivotally involved at the time, Wal-Mart's CFO became a key customer as P&G's objective became maximising Wal-Mart's internal profitability (Chandran 2003).

3 Integration of operations and processes

No evidence found in the material studied with reference to supply chain integration.

4 Planning and logistics control

In the early eighties, Wal-Mart placed heavy emphasis on developing and implementing tight supply chain solutions, catapulting the company to the top of the retail channel. Wal-Mart's supply chain technology has also allowed them to break the three-day barrier that some economists in the eighties felt was largely unbreakable. Wal-Mart is often able to replenish items on the Wal-Mart shelf in less than three days, not from the central warehouse to the shelf, but from the manufacturer to the shelf. With quick and reliable 2-day turnarounds, Wal-Mart is able to maintain lower levels of inventory and still meet customer demand. These lower inventory levels result in either a reduced floor plan with lower carrying costs and lower interest expense, or a greater diversity of products on the store shelves. Additionally, because Wal-Mart is better able to order inventory on demand, the company is in a better position to meet customer demand. Wal-Mart's supply chain technology allows the company to better avoid carrying an oversupply of fad items (Chandran 2003).

5 Quality management

No evidence was found in the material studied with reference to supply chain integration.

6 Information exchange

Wal-Mart and eight product manufacturers have begun testing electronic product codes, called EComputers, at selected centres in the US. Cases and pallets containing these products will feature EComputers when delivered to Wal-Mart's Sanger, Texas regional distribution centre. There, RFID readers installed at dock doors will automatically let Wal-Mart's operations and merchandising teams as well as suppliers know that this exact shipment of products has arrived and is inside the building. Cases will then be removed from pallets and processed as usual through the distribution centre. Wal-Mart is targeting 100 per cent readability of pallet tags through dock doors and 100 percent readability of case tags on distribution centre conveyor belts. The company has set a target for its top 100 suppliers to be placing RFID tags on cases and pallets destined for Wal-Mart stores, and believes that the implementation of this pilot scheme will pave the way for achieving this goal. Retailers such as Wal-Mart believe that the widespread implementation of RFID technology marks a sea change in the supply chain, much as the introduction of bar codes was as seen as revolutionary two decades ago. But while bar codes can tell a retailer that it has two boxes of a particular product, Wal-Mart's EComputers can help distinguish one box of that particular product from the next. This allows retailers greater visibility in monitoring product inventory from supplier to distribution centre to store (Chandran 2003).

7 Product development and design

No evidence was found in the material studied with reference to supply chain integration.

8 Market approach and marketing

No evidence was found in the material studied with reference to supply chain integration.

9 Cultural alignment

No evidence was found in the material studied with reference to supply chain integration.

10 Human resource management

No evidence was found in the material studied with reference to supply chain integration.

Table 21: Overview of applications of supply chain integration in the food and grocery sector, notably Wal-Mart

Factors of analysis	Description of application	Reference to 2nd order themes	Development level
Economic governance			
Integration of business activities	No evidence found in the material studied with reference to supply chain integration.	N/A	N/A
Partner sourcing and collaboration strategies	Engaging in strategic partnerships with big manufacturers of groceries in order to reduce cost levels, increase profitability and responsiveness	Strategic collaboration, commercial focus, market responsiveness	3. Closely connected stages
Production management			
Integration of operations and processes	No evidence found in the material studied with reference to supply chain integration.	N/A	N/A
Planning and logistics control	Planning and logistics are aimed at quick replenishment, avoiding oversupply and lowering inventory, supported by distribution centres, logistics service providers.	Inventory control, speed, focal firm dominance	4. Integrated supply chain
Quality management	No evidence found in the material studied with reference to supply chain integration.	N/A	N/A
Inter-firm governance			
Information exchange	Information is available in real time for collaborative forecasting planning and replenishment, using technologies such as RFID.	Real-time information, inventory control, market responsiveness	4. Integrated supply chain
Product development and design	No evidence found in the material studied with reference to supply chain integration.	N/A	N/A
Market approach and marketing	No evidence found in the material studied with reference to supply chain integration.	N/A	N/A
Social governance			
Cultural alignment	No evidence found in the material studied with reference to supply chain integration.	N/A	N/A
Human resource management	No evidence found in the material studied with reference to supply chain integration.	N/A	N/A

Table 22: Summary of applications of supply chain integration found in manufacturing per supply chain type

Types of supply chain structure	Type 1 supply chain: convergent assembly	Type 2 supply chain: divergent assembly	Type 3 supply chain: divergent differentiation
	Automotive Aerospace	Computers Large electronic systems	Clothing and textile Food and grocery
Economic governance			
Integration of business activities	• Integration of core capabilities in house, e.g. design, engineering. • Decentralised control of external activities via platform strategies, modular networks.	• Integration of sales and distribution by the manufacturer. • Integration is aimed at speed, fast channel management.	• Manufacturers rediscovered vertical integration, i.e. internalising business in house, to cope with volatility of markets, uncertainty of client behaviour.
Partner sourcing and collaboration strategies	• Close relations with suppliers. • Suppliers organised in a tiered structure. • Importance of first-tier suppliers e.g. in risk-sharing, cost cutting. • Reduction of supply base.	• Manufacturers enter in long-term contracts with key suppliers. • International sourcing has become important. • Modular approach to supply chains and production networks.	• Cost efficient procurement. • Purchasing is often organised globally to reduce costs. • Partnerships with suppliers to reduce cost levels, and increase profitability and responsiveness.
Production management			
Integration of operations and processes	• Suppliers integrated in order fulfilment process. • Integration of suppliers based on platform strategy, central control. • Suppliers treated as external business units.	• Integrate supply-assembly-sales • This includes suppliers connected to assembly. • Global activities also connected. • Integrated order and inventory management.	• Processes are highly integrated for speed, inventory control, and capturing customer information.
Planning and logistics control	• Assembly line is 'pulled' (kanban) and supplies delivered just-in-time (JIT). • Shorter lead-times by effective joint engineering with suppliers.	• Sales and resource planning for accurate forecasting and matching supply. • Large scale and global logistics. • Planning aimed at reduced lead times and time to market.	• Strict control of inventory for lowering inventory, preventing stock-outs or oversupply. • Quick replenishment supported by distribution centres, logistics service providers.
Quality management	• Suppliers responsible for the quality of own parts. • Strict rules and checks for quality assurance. • Supervision of suppliers' quality management systems. • Quality control while engineering	• High quality standards imposed onto suppliers. • Malfunctioning components require redevelopment. • Total quality management, continuous improvement to improve parts with suppliers.	• Quality management is routine. • Straightforward quality targets for suppliers to meet. • Quality of the end product and the design in the eyes of the end customer are key issues.

Inter-firm governance			
Information exchange	• RFID systems applied. • Network-structured knowledge-sharing, i.e. knowledge is property of the network. • Information exchange via supply chain information system, • Online collaborative engineering information environment. • RFID systems applied.	• Direct information passed from customer to assembly and suppliers. • Manufacturers dominant in information exchange. • Information is mainly exchanged to increase market responsiveness.	• Information exchange is real-time, supporting logistics and materials supply. • Reverse information to capture sales information, customers. • Supported by barcodes, microelectronics/RFID and e-commerce.
Product development and design	• Demand management, design for manufacture, capture design. • Manufacturer involves designers • Suppliers involved early in product development. • Prescribed engineering software central controlled engineering.	• Products are assembled to order based on customer demand. • Direct connection with customer to develop new products. • Manufacturer-lead designing. • reconfiguring existing products, decrease investments, lead time	• Manufacturer-controlled design. • Multi-functional teams including market experts and sales people with knowledge of new trends. • Postponement applied to assure products meet the latest trends and customer wishes.
Market approach and marketing	• Responsive order fulfilment. • Flexible forecasting of market developments into own process. • Anticipating future technology installed in new products.	• Direct sales to customers. • Products are assembled or engineered to order. • Aim is reduction of time to market and price.	• Maintaining customers' relation, being responsive, exclusive. • Releasing new products constantly for up-to-date image. • Adjusting products to regions.
Social governance			
Cultural alignment	• Supplier networks approached as families. • Supplier development programmes, supplier events. • Imposing standards onto supply chain partners.	• Supplier development programmes, supplier events. • Suppliers subscribing manufacturer codes. • Efforts aimed at building trust and achieving transparency.	• N/A
Human resource management	• Personnel exchange, joint teams. • Foster interaction and work exchange between individuals.	• N/A	• N/A

6.9 Reflection on key issues and contributions of this chapter to the thesis

The findings from the above explorations per industry sector and the example firms show considerable variation in the presence, appearance and level of supply chain integration practices (Table 22). The explorations and the examples are indicative for the understanding and the advancement of supply chain integration broadly across manufacturing industries. The findings indicate generally advanced forms and high levels of supply chain integration. The findings are based on secondary data and literature, and connect well to theoretical concepts related to supply chain integration as found in generic theory in the literature study of this thesis.

The explorations of the manufacturing industries in this chapter function as a bridge between the literature study and the building cases, and as comparisons for the building cases. The findings from these explorations are used to analyse the supply chain practices in building presented in the building case studies, and to support the relationship between theory and building practice leading to the eventual theory development of the thesis. In Chapter 8, the findings of the explorations will be further analysed, condensed and compared to the findings of the case studies in the building industry presented in Chapter 7.

7 Supply chain integration practices in the building industry

In this chapter, a selection of cases will be presented of firms in the building industry applying supply chain integration to their parts of the supply chain. The cases include firms in the positions of clients, developers, designers, builders and suppliers. The applications of supply chain integration vary from less advanced and 'partial' applications to more advanced and complete applications. In all cases, there is an increase in the level of integration and repetition of the product, process or inter-firm collaboration in the supply chain. For the exploration of the cases, the analytical model from Chapter 5 is used to describe the aspects of supply chain integration found in the cases. The case descriptions particularly include the procurement, development, design, production and supply phases of the building process. The exploration of the cases has been based on primary and secondary data, including interviews, reports and publicly available resources. An overview of the data gathered per case can be found in the appendices.

7.1 Outline of the building case studies

7.1.1 Cases selected for this thesis

The cases selected for this research were found by direct contact of the researcher with firms, and from a review of publicly available information, including profes-

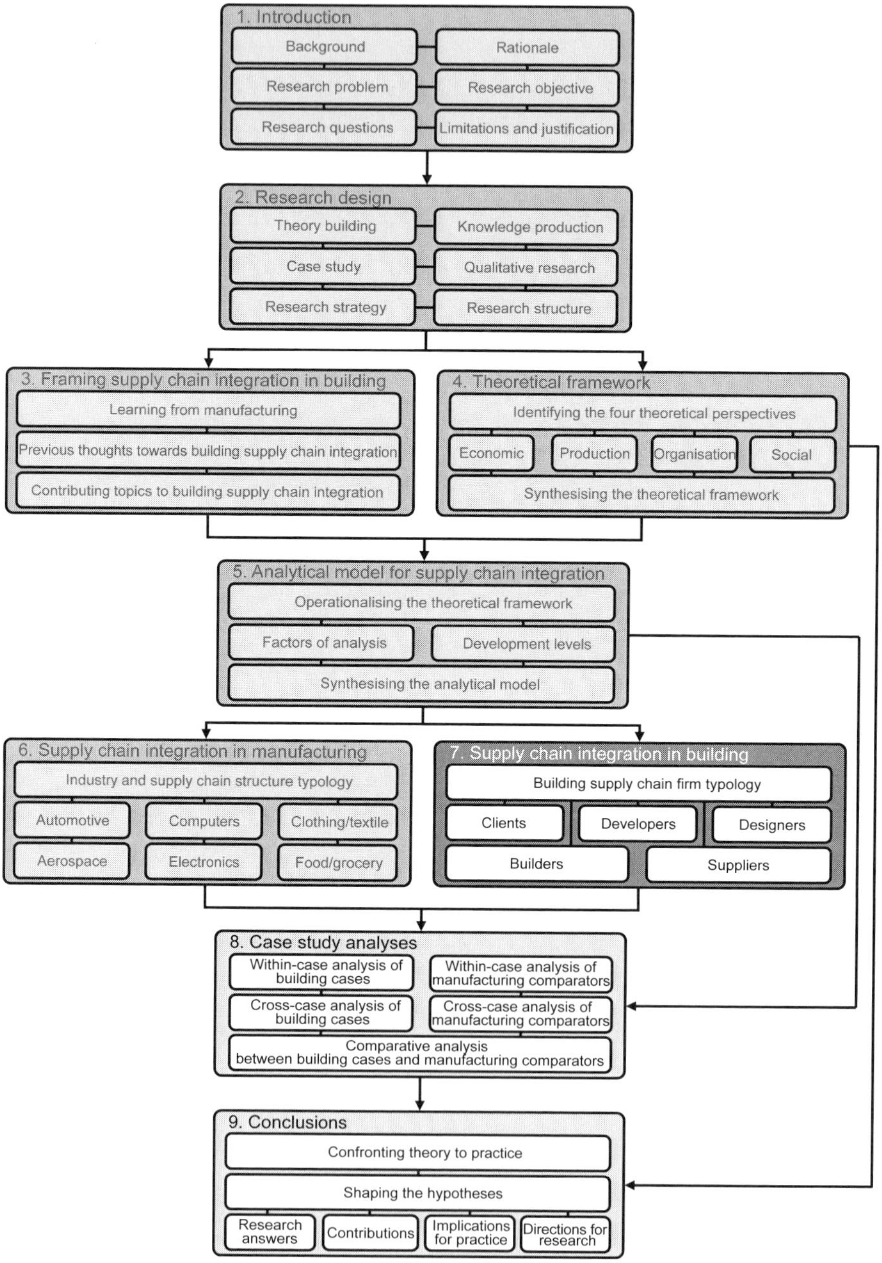
1. Introduction
Background
Rationale
Research problem
Research objective
Research questions
Limitations and justification
2. Research design
Theory building
Knowledge production
Case study
Qualitative research
Research strategy
Research structure
3. Framing supply chain integration in building
Learning from manufacturing
Previous thoughts towards building supply chain integration
Contributing topics to building supply chain integration
4. Theoretical framework
Identifying the four theoretical perspectives
Economic
Production
Organisation
Social
Synthesising the theoretical framework
5. Analytical model for supply chain integration
Operationalising the theoretical framework
Factors of analysis
Development levels
Synthesising the analytical model
6. Supply chain integration in manufacturing
Industry and supply chain structure typology
Automotive
Computers
Clothing/textile
Aerospace
Electronics
Food/grocery
7. Supply chain integration in building
Building supply chain firm typology
Clients
Developers
Designers
Builders
Suppliers
8. Case study analyses
Within-case analysis of building cases
Within-case analysis of manufacturing comparators
Cross-case analysis of building cases
Cross-case analysis of manufacturing comparators
Comparative analysis between building cases and manufacturing comparators
9. Conclusions
Confronting theory to practice
Shaping the hypotheses
Research answers
Contributions
Implications for practice
Directions for research

sional journals and Internet media. The cases selected have shown indications of applying supply chain integration to a certain extent, and of strategically addressing aspects of supply chain integration in their daily practice. For instance, some firms had engaged in strategic cooperation with other firms, or integrated preceding or subsequent business activities in the supply chain into their own organisation.

The cases apply to five firm types along the building supply chain: client organisation, project developer, designer, builder and supplier. This typology applies to the supply chain in commercial building and especially housing, which is the main focus area of this research. The cases studied in this research respectively include two housing corporations, a project developer, two architects, three builders, and a group of three suppliers. In a number of cases, one or two additional firms are involved in the case studies because they were involved in the supply chain integration strategy applied by the focal firm in that particular case (Figure 22). An overview of the data gathered per case can be found in the appendices.

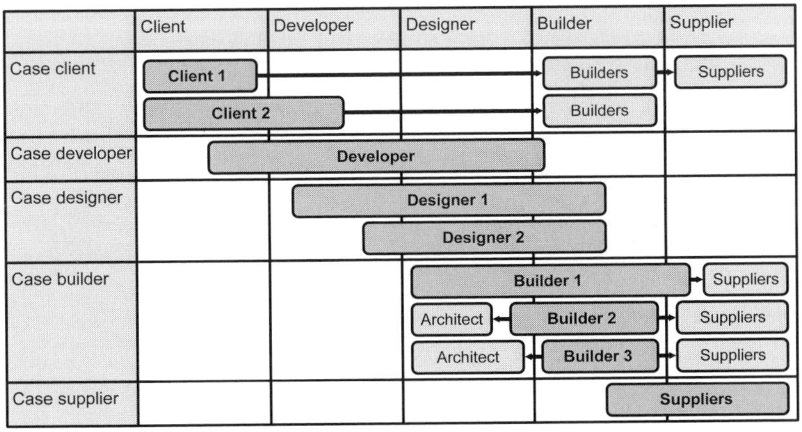

Figure 22: Overview of the building case studies

The cases collectively cover the different parts of the supply chain and theoretically represent a complete supply chain. Although the cases do not belong to one supply chain, analytically the findings from the different cases may be combined, and collectively they allow one to induce a comprehensive concept of supply chain integration for the whole supply chain. Although the unit of analysis is the focal firm, the supply chain integration strategies applied forward and backward in the supply chain conceptually integrate the whole supply chain.

7.1.2 Organisation of case descriptions

In each of the case studies, an introduction is given to the particular firm, and its supply chain strategy is observed, including a background, the objective and the firms involved in the supply chain integration approach. After the case description, we will further describe and explore the application of supply chain integration in the particular case following the ten factors of the analytical model. Each of the factors is described based on the research data. The research data consist mainly of interviews, documents from the cases, and writings about the cases. The interviews were organised, taken, transcribed and analysed based on the ten factors of analysis and the four levels of the analytical model. This has led to the descriptive summary below for each of the cases and for each of the ten factors, illustrated by representative quotations from the interviews.

Each of the case studies is concluded with an overview, including references made to the 2nd order themes i.e. codes used for the coding of the interviews (see Appendices), and an indication of the development levels of each of the factors of analysis. The different applications of supply chain integration in each of the case studies address the various factors of analysis differently, and to a variable extent. This depends on the firm type that is central in each case, i.e. client organisation, project developer, designer, builder or suppliers. This leads to a variable image of the application of supply chain integration across the case studies and poses a challenge to the further analysis of the outcomes. In order to facilitate comparative analysis in the following chapter, the overviews of the outcomes of the building cases have been presented in a similar manner as the manufacturing comparator studies presented in the previous chapter.

7.2 Supply chain integration practices applied by clients

7.2.1 Case of a housing corporation in operational cooperation with builders (Client 1)

In this case, the housing corporation has entered into a framework agreement with a number of contractors and maintenance firms for the project-based large-scale refurbishment as well as the regular maintenance of houses and apartments. The contractors involved had to organise the integration of specialist subcontractors and suppliers, and organise the involvement of residents and additional parties. The housing corporation had major overdue maintenance to perform on their housing stock. In order to resolve this problem, the corporation decided to revise the usual way of contracting building firms for those projects. They decided to contract a selected group of building firms for a longer period of time and develop cooperation in order to achieve higher levels of quality, improved coordination, and lower costs (Table 23).

The corporation owns a housing stock of 14,000 housing units. The corporation is committed to keeping the housing stock at an optimal level and to maximising efficiency. Supply chain integration has been embraced to improve the maintenance and renovation efforts in cooperation with selected building firms. The aim of the supply chain integration initiative by the corporation was to improve and sustain the cooperation with building and maintenance firms in the supply chain, and as a consequence to reduce inefficiencies and costs and improve quality and client satisfaction. The key factor has been to increase transparency and mutual trust. It has appeared to be difficult to actually quantify the improvements. In general, the quality has improved, coordination of work has improved, and as a result the cost levels are expected to be lower.

The corporation planned a series of refurbishment projects and invited a group of contractors, including their subcontractors and suppliers, to bid on and divide the work in mutual discussion. The projects were planned and coordinated completely by the contractors with their subcontractors and suppliers. The contractors were stimulated to solicit partners for long-lasting relationships for potential forthcoming projects. Part of the relationship was to apply lean construction principles to the coordination of the work aimed at the smooth and efficient execution of the work. The contractors were thereby challenged to apply innovative logistical methods. The contractors were also responsible for the communication with the architect and the residents. The corporation started the initiative with nine main contractors who had mobilised their own supply chains with specialists, and four maintenance firms, including a painter. Since then, the nine contractors have been reduced to six. In those specific cases, supply chain integration was applied in an initial series of refurbishment projects. Lessons were drawn from the projects for further application of supply chain integration.

The example project for this case study included the refurbishment of 114 apartments. The project consisted of the replacement of the entire front and rear façades of all apartments. The reasons for this operation were the bad condition of the window frames, low thermal insulation capacity of the façades, the presence of asbestos, and aesthetic reasons. These reasons were all the more important because the apartments were to be sold to the renters after refurbishment. To ensure a harmonious architecture for the neighbourhoods, the corporation decided to finish the replacement of the façades before the apartments were sold. The project was to apply supply chain integration between the corporation, main contractor, subcontractors and suppliers. The aim was to improve the price/quality ratio of the work, to improve the quality and thermal insulation of the apartments, and to decrease the inconvenience for residents during the work. The main contractor took the lead to coordinate the process from design and preparation to realisation. This contractor was also in charge of the communication with residents, the corporation, the architect and all other parties involved. In the design phase, the facades were designed in cooperation with all parties involved, including residents. The preparation of the work included the entire technical development of the façade, taking into account the residents' wishes and the additional aim of better thermal insulation. The façades needed to fit exactly to be placed onto the existing structure, which posed additional challenges with respect to the size, tolerance, installation and connections of the façade elements. In order to optimise the organisation of the site work, a test run was done for one apartment. For this test run, a detailed plan was drafted by the main contractor, the subcontractors and suppliers. Besides the site work, the communication and approach towards the residents were also incorporated into the planning, along with the measures for minimising the inconvenience caused for residents during the work. During the work, frequent meetings assured the progress of the work and the communication with residents, who were involved in a few of the meetings.

1 Integration of business activities

The corporation did not internalise any business activities. Instead, the corporation gave contractors a larger role in the procurement, further subcontracting, project management and communication with residents during the refurbishment. Because of this, the corporation was able to economise on internal efforts with regards to the procurement, project management and organisation of the communication with residents. An additional outcome was the association and mutual coordination among the contractors, motivating them to take up lean construction as a standard means of project management with subcontractors.

Corporation: 'We wanted the contractors to take on more responsibility and smooth the procurement process. We knew we would never get the exact price beforehand. Therefore, we said goodbye to the traditional procurement methods of open tendering with briefs and drawings and putting those on the market. Our new approach has been to invite a group of contractors and let them divide the work amongst them and fix a set of prices everyone can live with. We are convinced this is the way forward for the good of all parties'.

2 Partner sourcing and collaboration strategies

Initially, the supply chain procurement included framework agreements with four maintenance firms for regular maintenance and small refurbishments, and nine contractors, including a total of thirty specialist subs for project-based, large-scale refurbishments. The procurement process was organised in a collective format with all contractors mutually selecting the work each contractor was able and willing to do. In this collective process, all contractors were able to discuss their preferences and the capacities they had available for the projects. The procurement was based on prefixed element prices and a model contract. A major contract condition was to share operational benefits with the corporation. Another condition was that the contractors had to organise all subcontracting to specialists and suppliers within the set of prefixed prices. The subcontracting and approach to specialists and suppliers, however, has generally been traditional.

Contractor: 'We were able to pick the work we knew we were good at, knowing what work the other contractors had picked. During the collective procurement sessions some contractors posted their names for all the work. The others knew they could not do it all, because of their capacity and expertise, so they were corrected by the others.'

Supplier: 'In theory, we should have been part of the supply chain integration. However, in practice we were contracted traditionally. The contractors have a framework agreement, but not the subcontractors and suppliers. The corporation would benefit if they would also pool all subcontractors and suppliers, to be selected from by the contractors. This would benefit all, since now we are not sure whether we will be selected at all.'

3 Integration of operations and processes

Besides the procurement process and the project management, the corporation and the contractors have not integrated any further processes. Within the project, the main contractor and the subcontractors and suppliers have intensively coordinated technical engineering solutions and drawings, and the planning and organisation of mutual operations on site, based on the concept of lean construction. However, actual processes and operations have remained separate.

Contractor: 'For us, the method of lean construction has been a great advantage for the organisation on site with our own people and the subcontractors. The work crews have communicated intensively about the lean planning. But for the rest, operations have remained separate.'

4 Planning and logistics control

Planning and logistics were based on the lean construction approach, i.e. joint planning of the work by the contractor and subcontractors, followed by intensified conversation about the planning during the work in order to accomplish the planning. Further logistics of bringing materials to the site were closely discussed by the contractor with the suppliers, to be able to smoothly include the installation of elements into the planning, e.g. the replacement and installation of integrated façade elements.

Supplier: 'We have discussed and communicated closely with the contractors about the exact date and time, when and where to deliver each façade element in order to keep operations flowing.'

5 Quality management

Quality control was based on the quality management systems of the subsequent parties. There was no joint quality system in place. As a side effect of the intensified collaboration between contractor and supplier, including during the engineering, the quality improved considerably. For instance, the insulation figure of the façade elements went up and in fact outperformed the terms of condition set by the design.

Contractor: 'Without explicitly aiming for it, the façade elements apparently insulated the apartment better than expected. After installation and finishing the project, the thermal measures taken in the apartments showed the elements outperformed, shifting the apartments' thermal performance one thermal class up. This was a great surprise to everyone.'

6 Information exchange

Information was shared in a traditional way. There was no joint information system in place. Separate project files were stored at different locations. Collective digital facilities to store and share documentation were not used.

Contractors: 'Since everyone worked closely together, communication was good, and apparently we didn't need an additional solution for information exchange.'

7 Product development and design

The initial architectural design of each project was drawn by separate architects using their own design formats. For each project, the architectural design was translated into a technical design in collaboration with the contractor and the specialist subcontractors and suppliers. In quite a few cases, the architect was involved in the technical engineering and changing the architectural designs based on this feedback.

Contractor: 'The architects had designed the façades, but new insights during the engineering and detailing, mainly on the part of the supplier, made it necessary and beneficial to change the architectural design based on that. The architect was cooperative in all this.'

8 Market approach and marketing

In each of the projects, the residents already living in the houses were intensively involved in the planning and the execution of the project. For each project, one resident was the spokesperson and had to be involved directly by the contractor. After the project's completion, residents were asked their opinion about the result and the process. This opinion was crucial in the evaluation of each contractor, and actually led after the first round of projects to the reduction of the number of contractors from nine to seven for the large-scale refurbishments and for the second round.

Contractor: 'We had to get used to communicating directly to the residents, and allowing them to be a part of the project. In the end, it turned out well, and the residents were generally happy, too.'

9 Cultural alignment

The corporation attached great value to the common understanding of the philosophy and the benefits of supply chain integration by contractors, subs and residents, and beyond. For instance, a steering committee was installed to supervise the supply chain integration, and contractors were often brought together in joint meetings and involved in joint presentations, for instance, at national symposia with contractors, subcontractors and even residents.

Table 23: Case overview of the housing corporation in operational cooperation with builders (Client 1)

Factors of analysis	Description of application	Reference to 2nd order themes	Development level
Economic governance			
Integration of business activities	The corporation did not internalise business activities. It took al lead over contractors and subcontractors. At the same time the clients reduced its own effort e.g. for procurement.	Focal firm dominance, strategic cooperation,	1 .Independent stages
Partner sourcing and collaboration strategies	Framework agreements for maintenance and small refurbishments with four maintenance firms, and initially nine contractors, including a total of 30 specialist subs for large refurbishments.	Strategic cooperation	3. Closely connected stages
Production management			
Integration of operations and processes	Besides the procurement, the corporation and the main contractors have not integrated any processes.	Strategic cooperation	2. Loosely coupled stages
Planning and logistics control	Planning and logistics were based on a lean construction approach. Logistics to site were closely discussed with the contractor and suppliers.	Speed	3. Closely connected stages
Quality management	There was no joint quality management.	Specialist roles	1. Independent stages
Inter-firm governance			
Information exchange	There was no joint information sharing.	Specialist roles	1. Independent stages
Product development and design	Architectural design was translated into a technical design in collaboration with the contractor and the specialist subs.	Teamwork	2. Loosely coupled stages
Market approach and marketing	Residents were intensively involved in the planning and execution of projects. One resident per project was the spokesperson for the contractor. After the projects, residents were involved in the evaluation of each contractor.	Market responsiveness	2. Loosely coupled stages
Social governance			
Cultural alignment	Steering committee installed to supervise the supply chain integration, contractors often gathered in joint meetings, and joint presentations at symposia involving contractors, subs and residents.	Strategic cooperation, teamwork	2. Loosely coupled stages
Human resource management	Establishment of an open relation informally between representatives of the client organisation and the contractors.	Teamwork	2. Loosely coupled stages

Corporation: 'We want contractors and residents alike to be part of this movement. We attach great value to spreading this idea as broadly as possible, with the help of everyone. This philosophy of supply chain integration needs to spread in our sector unchecked, like an oil slick on water.'

10 Human resource management

In the collective procurement sessions and joint meetings with the contractors, many members of the corporation's staff were involved, e.g. the buyers and project managers. The aim was to establish an open relation informally between representatives of the client organisation and the contractors. With the subcontractors and suppliers, however, the interaction on the level of personnel has been thin.

Contractor: 'In a few instances, the personal collectivisation among the corporation and the collective contractors was quite intensive. However, the meetings are quite functional in a way. We would welcome more such meetings to exchange knowledge and experience with the new way of working among the contractors.'

7.2.2 Case of a housing corporation in legal partnership with builders (Client 2)

In this case, in 2003 the housing corporation has entered into a 15-year Public Private Partnership (PPP) with a municipality, and into a legal partnership with four contractors. The partnerships are all aimed at realising a budget-neutral programme of multiple housing developments and redevelopments. The programme is organised in smaller tracks of selected projects that need to result in a neutral investment. Each project is executed by one or more of the contractors (Table 24).

The order and priority of the projects therefore revolved around financial rather than political considerations. The corporation launched the initiative to proactively plan the developments in accordance with the municipality. The municipality was encountering a large backlog in the redevelopment of old neighbourhoods and foresaw the need to develop new housing areas in the near future. The issue was how to finance this. The corporation took up the challenge and mobilised four contractors who were willing to step in and had previously formed a legal entity.

The aim of the multi-project approach is to have built 750 units by 2015 with a zero-profit total result. Projects with a negative result are compensated for by projects with a positive result. Projects are planned in sequence and calculated financially by means of a collaborative calculation method in which all cost indicators and financial parameters of the collective parties are unified. Projects are then grouped with a total break-even financial result of the projects together. Each group of projects is then commissioned to the contractors in mutual agreement based on the main planning and financial scheme as agreed before.

The corporation leads the collaboration, including the municipality and the four contractors involved. The entire operation is formalised in a joint legal organisation structure including municipality, corporation and contractors. The contractors are collaborating within a legal entity. The contractors plus the corporation form a higher tier legal entity. This entity plus the municipality form the overarching legal entity.

The example project for this case study included 48 newly built houses in the sales segment. The project was built by a combination of three of the consortium builders for the houses and for the civil works. The project was carried out in the period 2005-2007 at a total budget of € 12.600.000.

1 Integration of business activities

The supply chain integration takes place particularly on an institutional level. The corporation is a shareholder in the joint legal partnership, together with the four builders. In addition, this partnership is a shareholder together with the municipality in a joint Public Private Partnership. As part of this, the corporation has the authority to plan and initiate the projects according to the priority that the private partners, i.e. the corporation and the contractors, have set.

Corporation: 'We took the initiative to plan the developments around the city that we thought were necessary. The municipality relinquished a leading role in this process.'

2 Partner sourcing and collaboration strategies

The four contractors had previously taken part in a joint legal undertaking. This entity was invited by the corporation, and after a selection procedure became a shareholder in the larger legal entity together with the corporation. Based on a framework agreement, projects are being planned in budget-neutral groups, and based on the planning, the projects are procured and built by one of the contractors.

Contractor A: 'The procurement system is fairly simple. The client prioritises the developments to be undertaken, and then together we develop those in such a way that a budget-neutral solution results.'

3 Integration of operations and processes

The boards of the respective joint partnerships are staffed jointly by management representatives of the respective organisations. Besides the planning, development and procurement of the projects take place in a joint effort with the municipality, the corporation and the builders. On an operational level, the project management is a joint effort by the corporation and the contractor building the project. Further down the supply chain, however, no integration of processes takes place.

Contractor B: 'We are responsible for the civil works of every project. However, we are not involved in the development of those projects. We have no say in that, nor are we involved in the planning of the project. So in theory, we are part of the joint partnership, but in reality, for us it comes down to a fairly traditional project.'

4 Planning and logistics control

On a partnership level, the planning of the projects is a joint development and financial effort with the aim to be budget neutral. But on an operational level per project, the planning and logistics are organised traditionally by the contractors. The corporation does not interfere with the execution of the projects, nor are there clear or shared operational benefits that could result from repetition. The collaboration does not aim for such refinements.

The corporation: 'In fact, we have failed in making clear what benefits we want from the partnership with the contractors. The contractors get their work rather easily. Besides developing the projects in a budget-neutral fashion, there are no improvement targets or financial incentives for the contractors to improve based on the repetition. And thus there is no benefit there for us. This is actually an omission in the whole deal.'

5 Quality management

Quality control is based on the quality management systems of the subsequent parties. There is no joint quality system in place. The quality of the developments and the building of the projects as well as the 'financial quality' are secured on a high aggregation level by the joint development of the projects by representatives of the client organisation and the contractors.

Contractor A: 'To the projects we apply our own quality regime. The advantage is that we have developed the projects jointly with the client, so after that we can focus on getting the quality we jointly defined.'

6 Information exchange

Information is shared in a traditional way. There is no joint information system in place. The level of information sharing in the joint development phase is higher than afterwards during the further design and construction, which are accomplished in a traditional way.

Contractor A: 'We get the information straight from the client. After the joint development we take over till the delivery of the project back to the client.'

7 Product development and design

Projects are developed collaboratively, however, per individual project, thus not in a multi-project approach to each of the budget-neutral tracks of the projects, for instance. In fact, the projects per track are of different kinds (different markets, quality levels, buy vs. rent) to be able to find the right mix of projects that make a budget-neutral track. This hinders such possibilities as working with product templates, for instance. There are no joint product development or design principles in place between the respective organisations. Architects are invited per project and are thus not involved on a multi-project basis.

Corporation: 'Joint project development is the key to achieving successful projects and the budget-neutral tracks. We need to balance the tracks and projects to achieve those goals. So we can't really work with pre-developed products.'

8 Market approach and marketing

The end client, one of the buyers or residents of the houses developed per project, is involved in a traditional way. Projects are marketed in a traditional way. Rentals and sales are traditionally organised through the client organisation.

Corporation: 'The partnership is basically aimed at getting the work done with the contractors at an efficient price level without financial surprises. The client side of the business has never been a goal of the partnership as such.'

9 Cultural alignment

Besides the existence of the legal partnerships and the joint publicity given to those partnerships, there is no further alignment between the respective organisations. On a management level, the municipality, the corporation and the contractors form joint boards of management that must ensure the alignment of the whole undertaking.

Contractor A: 'We come together with the boards of the legal partnerships regularly to discuss the collaboration in general, financial issues, and any possible problems that may have occurred. Beyond that, the respective organisations need to follow what has been agreed on by the boards.'

10 Human resource management

Besides the joint representation of the organisations involved on a management level in the boards, on all the projects there are always pairs of people responsible for the project management. Each pair consists of one person from the corporation and one who acts on behalf of the contractors. The joint effort is particularly aimed at the joint development, the design and the preparation phases. During construction and up to the delivery, the project manager from the corporation is involved in the background.

Corporation: 'We have a project manager on all projects together with a project manager from the contractors. It is always a pair of people. Particularly in the first phases of each project we are involved. During construction we are informed and involved when needed.'

Table 24: Case overview of the housing corporation in legal partnership with builders (Client 2)

Factors of analysis	Description of application	Reference to 2nd order themes	Development level
Economic governance			
Integration of business activities	Supply chain integration takes place particularly on an institutional level. The respective organisations are shareholders in joint legal partnerships.	Strategic cooperation	4. Integrated supply chain
Partner sourcing and collaboration strategies	Based on the framework agreements of the legal partnerships, series of projects are planned in budget-neutral tracks, based on the planning procured per project.	Strategic cooperation	4. Integrated supply chain
Production management			
Integration of operations and processes	Planning and the procurement of the projects take place in a joint effort with the municipality, the corporation and the builders. On an operational level, no integration of processes takes place.	Teamwork	2. Loosely coupled stages
Planning and logistics control	On an operational level, planning and logistics are organised traditionally per project by the builders.	Focal firm dominance, specialist roles	1. Independent stages
Quality management	There is no joint quality management system.	Specialist roles	1. Independent stages
Inter-firm governance			
Information exchange	There is no joint information sharing system on operational level. Only incidental information shared on management level.	Strategic cooperation	1. Independent stages
Product development and design	Projects are developed collaboratively per individual project. Architects are invited per project. No multi-project approach to the tracks of projects.	Teamwork	2. Loosely coupled stages
Market approach and marketing	Involvement of the end client and marketing efforts are no part of a joint approach, organised in a traditional way.	Specialist roles	1. Independent stages
Social governance			
Cultural alignment	Besides the joint legal partnerships and joint publicity, no further alignment between the respective organisations.	Strategic cooperation	1. Independent stages
Human resource management	Management of the legal partnerships and the projects provided by representatives from the respective organisations.	Teamwork	3. Closely connected stages

7.3 Supply chain integration practices applied by developers

7.3.1 Case of a full-service project developer (Developer)

In this case, the project developer has developed a business model of internalisation. Internalisation takes place in two directions. First, the internalisation refers to supply chain activities such as in-house conceptual design and management contracting. Further, internalisation takes place over the life cycle, such as delivering security or energy with the building delivered. Originally, the developer was an international company and started years ago with a few projects in the Netherlands. From the first projects done in the Netherlands, the developer applied a full-service approach to the built asset and the accommodation delivered to the user. Since this start in the Netherlands, the developer has held on to its philosophy and applied it to all of its subsequent projects. This has mainly consisted of commercial and utility buildings, such of offices and business parks in the Netherlands and abroad (Table 25).

The developer wants to be its own client and supply the built asset, or rather, the full-serviced accommodation to the users. Ideally, the firm develops the product first and then searches for a client for it. The developer tries to stay as long as needed in the facility management of a built asset delivered, for instance, 10 years or even longer. The developer's approach is aimed at developing and building real estate for itself as the client, and then offer that to end users, e.g. an international firm looking for accommodation in the Netherlands. As part of this approach, the developer also takes care of the maintenance and the facility management, i.e. operating the building, including security services and energy supply. In the developer's view, users should not have to bother and must experience 'full service'. The user comes first in the approach, then the product, and finally the location. For each project, the developer does a functional analysis and sets the branding of the product to be built without the help of an architect. Based on that, the developer later invites an architect for the actual design.

For the interior design, the developer has a partnership with an interior architect who fully understands the developer's philosophy. At one point, the developer had taken over an architecture firm, but this has since been dissolved. For each project, the developer discusses the building process with the contractor. If, for instance, the construction period would take too long, TCN demands creativity on the part of the builder to shorten that period and to resolve issues such as capacity or external factors. In some cases the developer acts as a management contractor and subcontracts the work to subcontractors. Previous plans were devised to take over a contractor or close a to deal with a fixed contractor, but these plans also did not work out. Central to the potential takeovers was the question of whether it would generate added value for future projects in the long term. In addition to this, the developer does work in integrated teams with building partners, taking part in private contracts only, not in open tenders.

The example project for this case study included the redevelopment of an existing commercial area. This area was acquired by the developer and then redeveloped for the existing users. The developer not only redeveloped and built new buildings in the area, but also took over all facilities for the area, including the energy delivery and exploitation of the energy infrastructure.

1 Integration of business activities

The developer's approach to supply chain integration is aimed at integration or dominant coordination of the life cycle from land acquisition to facility management of the built facility for the user, including all intermediate phases. The developer prefers to be its own client for the built facilities and then delivers the built facility and full service along with it to the user. This may vary per project, for instance, delivering security or energy supply with the building.

Developer: 'We want the user to experience full service and have nothing to worry about. That means we want to deliver a built facility and everything that comes with it. We must therefore fully understand the user's wishes and translate that to every aspect of the built asset. That makes it necessary to be in control of the entire process up to the delivery.'

2 Partner sourcing and collaboration strategies

Based on the need to control the entire process, the developer tries to fix relationships wherever it is possible and effective to team up with architects, builders, energy suppliers and other firms. In a few cases, this is multi-project, but often this remains one-off. For instance, with energy suppliers, the developer tries to settle long-term deals to deliver energy to multiple assets. However, in the case of architects, the developer selects architects based on the types of project.

Developer: 'For commodities that we need for all of our assets, such as energy, we try to fix deals with energy suppliers, for instance. We have tried this with contractors and architects, too, even trying to take them in-house. However, this didn't work out well.'

3 Integration of operations and processes

For its projects, the developer coordinates all processes linked to the entire real estate process, including the design and building done by external parties. In a few instances, the developer internalises operations and processes such as the concept development and the functional design, which are done before the architect starts designing. With regard to the actual building, the developer in a few cases acts as management contractor rather than hiring a main contractor. Moreover, the interior design and fit-out of buildings is often partly done in-house, at least as a start before external parties are hired to do such jobs.

Developer: 'All parties we work with need to understand our philosophy. If we can't find the right party to do it for us, we need to do it ourselves. Although the temptation is strong to simply take activities in-house, which we actually tried a few times, we aim for finding the right set of partners for our projects'.

4 Planning and logistics control

Following the integration of operations and processes, the general overall project planning and the delivery of the asset to the user is coordinated by the developer. Further work planning and logistics on site are generally coordinated by the contractor of the project. However, the developer maintains control of the entire project, including during building until project delivery and handover of the building to the user, even including all the after-sales and facility management that comes afterwards.

Developer: 'As part of the full control we want and need to exercise over projects, we also control the execution of the project. If we act on a project as management contractor, we will do the planning and logistics ourselves with the subs and suppliers to the project. If we hire a main contractor, he will organise all this.'

5 Quality management

Although quality standards are high, there is no integrated quality management system in place to control the total quality, including the quality delivered by third parties. The strict control of the entire process, however, also addresses quality, so implicitly, there is a dispersed total quality management system.

Developer: 'Since we are often our own client and responsible for the proper functioning and the life cycle costs of the built assets we deliver to the user, we have great interest in high quality. So this is a major aspect throughout the entire process, and thus also a major point of interest of our project management.'

6 Information exchange

The developer has no integrated information management system in place. The information exchange on projects is coordinated by the developer in a rather traditional way. The developer controls the major information flows.

Developer: 'We want to manage the essential core of our projects, and a few crucial details. The rest is done by the project partners hired.'

7 Product development and design

Product and concept development are a key issue for the developer and takes place internally. This occurs at the very beginning, or preferably before a project is started. It sets out the product to be developed for a user. The concept development of the product is essential to its success. Once the developer is comfortable with the concept, external designers generally develop the architecture and the internal design of the building.

Table 25: Case overview of the full-service project developer (Developer)

Factors of analysis	Description of application	Reference to 2nd order themes	Development level
Economic governance			
Integration of business activities	Integration or dominant coordination of the life cycle from land acquisition to facility management including all intermediate phases. Preferably the developer is its own client.	Focal firm dominance, absolute control, market push	4. Integrated supply chain
Partner sourcing and collaboration strategies	Fixed relationships when possible with architects, builders, energy suppliers (for facility management).	Commercial focus	2. Loosely coupled stages
Production management			
Integration of operations and processes	The developer coordinates all processes linked to the real estate process, including the design and building, done by external partners.	Focal firm dominance	2. Loosely coupled stages
Planning and logistics control	General project planning is coordinated by the developer. Work planning and logistics are coordinated by the builder of the project.	Specialist roles	2. Loosely coupled stages
Quality management	No formalised integrated quality management system in place.	Focal firm dominance	1. Independent stages
Inter-firm governance			
Information exchange	No joint information management system in place.	Focal firm dominance	1. Independent stages
Product development and design	Product development and concept design takes place in house. For the architecture and internal design, designers are contracted.	Focal firm dominance	4. Integrated supply chain
Market approach and marketing	The developer tries to be its own client as far as possible and to deliver the building, including the total facility management, to the end user.	Market responsiveness, market push	4. Integrated supply chain
Social governance			
Cultural alignment	The developer aims to stamp its principles onto all its projects and the supply chain; however, no active alignment takes place with building partners.	Focal firm dominance, commercial focus	2. Loosely coupled stages
Human resource management	Personnel in the supply chain stay part of their own organisation	Commercial focus	1. Independent stages

Developer: 'The one thing we really want and need to do in-house is the concept design. This needs to completely correspond with our philosophy, and this is crucial to the product success. So we need to be on top of this.'

8 Market approach and marketing

The developer tries to be its own client as far as possible, and to deliver the building, including the total facility management, to the end user, e.g. a firm seeking accommodation, including all the services that accompany it. The developer sells its buildings as products, and branding is essential. All of the developer's buildings are recognisable as such with their trademark displayed clearly on it.

Developer: 'Our clients don't buy or rent a building; they acquire a trademarked product with all the services they want and all the guarantees you may expect from it.'

9 Cultural alignment

The developer aims to stamp its principles onto all its projects, buildings delivered and the supply chain. However, cultural alignment is strongly applied internally within the developer, rather than externally upon supply chain partners. No active alignment takes place with regard to external building partners.

Developer: 'This firm thrives on a strong corporate culture, flat hierarchy, and a few very clear and strictly applied rules. That's a thing everyone here knows and lives up to.'

10 Human resource management

Although the developer's control over the supply chain for its projects is strict and characterised by comprehensive integration, it extends to the level of the firms in the supply chain. Personnel of those firms remain part of their own organisation and are not pooled in a joint approach.

Developer: 'In our projects we generally chair all meetings, often also construction meetings during execution. This is where personnel from the other firms are pooled together rather traditionally.'

7.4 Supply chain integration practices applied by designers

7.4.1 Case of a developing and contracting architect (Designer 1)

In this case, the architect has internalised a developing and a management contracting branch within his own firm. The architect has an established reputation for designer-led procurement and management of building projects. The firm has long been a dominant and trendsetting architecture firm with multiple disciplines in house, and thus able to take on more activities outside the mere design of buildings. Development, design and management contracting take place preferably in-house. However, only in a few cases are all three disciplines applied in one project. In all cases, design is leading the project development as well as the contracting (Table 26).

The architect has a preference for industrial architecture, prefab building, and modular design. The firm is known, for instance, for the industrial design of buildings. In this area, the architect had always been active in product development of building parts with specialist suppliers, e.g. for the development of system façades and climate floors, and always with an industrial appearance.

The architect aims to be a dominant player in the high end of the architecture sector and sustain the branding of the typical industrial-looking designs. In order to achieve these objectives and the high standards of the architecture, the architect has decided to bring project development and contracting under the same roof. Being able to coordinate and manage the subsequent phases of the design enables the architect to apply his design philosophy throughout the supply chain.

In addition to the integrated approach to development-design-build, quality and efficiency are achieved by consistent application of design details and components, and employment of a small number of subcontractors with whom the architect has long-standing relationships. Besides the projects, the longer-term product development of building parts is engaged in with specialist suppliers, whom the architect tries to use in further projects.

In the period 1997-2003, the architect designed and built a series of what was supposed to be 20 postal distribution logistics centres around the country, for a total budget of approximately 14 million euros. Due to changes in the postal market, fewer centres were actually built than planned. The design was modular and could be adjusted to local situations and capacity needs, e.g. the location and size of the loading platform for the post to be loaded and emptied from the trucks into the centre and vice versa. The product was developed together with a few materials suppliers, e.g. steel structure, façade elements and roofing elements.

1 Integration of business activities

The architect has integrated project development and contracting within his own organisation. This approach is rather unusual in the architecture business but has been rather effective in accomplishing good projects efficiently. It has required the architect to take different disciplines on board his own organisation.

Architect: 'The integrated approach has developed rather naturally. The existing organisation has not changed much. The additional activities of the development and contracting have been put in separate legal business entities.'

2 Partner sourcing and collaboration strategies

Partner selection takes place per project. This also applies to the internal business units. However, certain subcontractors have a long-standing relationship with the architect and are often involved in his projects. For the product development, such as integrated climate façade and floor systems, the architect has established development partnerships with specialist suppliers.

Architect: 'Generally we contract subcontractors per project. However, for certain building parts we have had good experiences with certain subcontractors who, because of our long-standing relationship, are given priority for every new project. The continuation of the relationship depends on the quality delivered by the particular subcontractors.'

3 Integration of operations and processes

Within the offices of the architect, the developer, the designer and the management contractor are legally separate business units. There are no operational links between them, nor is there integration with processes of other parties in the supply chain.

Architect: 'The integration we apply merely has to do with latent links that we can install when appropriate for a project. In such a case, for each project we can easily link processes in the way the project requires.'

4 Planning and logistics control

The project planning is coordinated from the design, and is passed on via the in-house contractor to the external contractor. The in-house contractor coordinates the process up to the site logistics. The external contractor does the actual building on site.

Architect: 'We have our own construction professionals who act as project managers for the construction on site, together with the management of the main contractor or subcontractors whom we contracted to actually do the building and installation on site. We do the work up to the preparations on site, say, till putting up the fences around the site.'

5 Quality management

No integrated quality management system has been put in place. The quality is managed from the design and from pooling the internal business units and the contractor or subcontractors for each work. The strong dominance of the process and the supply chain from the design, however, implies an integrated approach to quality control.

Architect: 'All standards and controls for all our work, including its quality, refer back to the design. For us as an architect, the design is the central point from which we coordinate the rest of the business. The quality of the design and its realisation is a major issue here.'

6 Information exchange

No joint information management system has been put in place. This applies to the internal business units as well as the external project partners with whom the architect works.

Architect: 'We actually have no joint place for information sharing besides our own digital server. This is just used to store documents. There is no document management system among the different units, or with the project partners. This is resolved traditionally.'

Table 26: Case overview of the developing and contracting architect (Designer 1)

Factors of analysis	Description of application	Reference to 2nd order themes	Development level
Economic governance			
Integration of business activities	Integration of project developer and contractor within the organisation.	Absolute control	4. Integrated supply chain
Partner sourcing and collaboration strategies	Partner selection per project. Parts development in partnership with specialist suppliers e.g. climate floor system.	Strategic cooperation	2. Loosely coupled stages
Production management			
Integration of operations and processes	Within the offices of the architect, the developer, designer and contractor are separate units. There are no operational links with processes of other parties in the supply chain.	Specialist roles	2. Loosely coupled stages
Planning and logistics control	The project planning is coordinated from the design and is passed on via the in-house contractor to the external contractor. The in-house contractor coordinates logistics up to the site.	Absolute control	2. Loosely coupled stages
Quality management	No formal quality management system in place.	Specialist roles	1. Independent stages
Inter-firm governance			
Information exchange	No joint information management system in place.	Specialist roles	1. Independent stages
Product development and design	The architect coordinates the design and the parts development. The design and the parts have typical features, i.e. industrial intelligent design of parts.	Design rationality, technological supremacy	2. Loosely coupled stages
Market approach and marketing	Most projects approached as one-off, some as a package, e.g. postal distribution centres. All projects are marked with the typical industrial look.	Technological supremacy	2. Loosely coupled stages
Social governance			
Cultural alignment	No active alignment with building partners	Specialist roles	1. Independent stages
Human resource management	No joint HR policy in place	Specialist roles	1. Independent stages

7 Product development and design

For the design of projects, the architect aims to align the design with the typical industrial-looking branding of all its designs. The resulting design of each project has the repetitive typical features, e.g. steel and glass façade elements. In addition, the architect has initiated long-term product development of building parts such as insulated steel façade elements and integrated climate floor slabs, all in accordance with the industrial branding.

Architect: 'Besides the application of new technology into our projects, we actively do product development notably for the façades and floors. For the technical part we collaborate with specialist suppliers, for instance for the insulation of the steel façade elements, and the installations within the integrated steel floor slabs we developed.'

8 Market approach and marketing

Most projects are approached as a one-off; however some, such as the logistics centres, are approached as a multiple package deal. Although most projects are generally approached as one-offs, all designs are marketed as typical products by the architect, i.e. industrial-looking buildings.

Architect: 'Our designs are typically industrial-looking. This is our branding that we carefully apply to all our designs and product development. This is our philosophy that we strongly hold fast to and strive to develop further in every subsequent project.'

9 Cultural alignment

Besides the central coordination of activities, no active alignment takes place between the internal business units, nor with other building partners with regards to work methods or cultures.

10 Human resource management

Within the firm, people from the different business units work jointly on projects if the project is a combination of development-design-contracting. However, there is no policy for this. Outside the firm, no joint approach to personnel exists, either.

7.4.2 Case of an architectural engineer (Designer 2)

In this case, an architectural engineering firm has internally applied Building Information Modelling (BIM) as a central way of working. The engineer acts as an intermediate between architecture and engineering. The designing and engineering is done in house. The firm includes a part of the project development if required, and has also internalised a management contractor. The approach is fully based on BIM. Often clients involve the firm if they want the project in BIM; sometimes the architect does to have a design translated into a 3D model, or the contractor calls for this when a project is too complex to translate it technically. The objective of the firm is to play the role of systems integrator via BIM. Not only technically, but also organisationally in projects. This requires a position in the early stages of projects. During construction, the firm aims to coordinate the construction based on their BIM, as a management contractor (Table 27).

In general, the firm aims to discuss a central role with the client in which they can coordinate the other parties including the architect. But often the firm is hired by an architect or even a contractor to technically solve complex designs. For the design of a new complex steel structure for the large-scale refurbishment of an existing monumental building, the steel supplier was involved in the engineering. The collaboration was particularly intense for the integration of the design of the steel structure, including the exchange and integration of the BIM models of the whole building and the steel structure. The firm also has development partnerships in place with other parties, such as a supplier of hollow floor slabs, with whom the firm has developed a specific floor system to be used in projects.

1 Integration of business activities

The engineer has in-house engineering expertise, BIM design facilities and expertise, and an in-house management contractor. The engineering and BIM designing in particular are integrated in all projects the engineer does.

Engineer: 'We start all projects in BIM from the outset and approach the project both from a design and an engineering angle. We have the expertise to do both, which we do. And if we get the chance, we integrate the contracting, too.'

2 Partner sourcing and collaboration strategies

Partner selection takes place per project. This depends in the way the engineer is involved in the project, whether it be by the client, architect or contractors. In the first two cases, the engineer is often involved in selecting the subcontractors and suppliers. With a few suppliers the engineer has entered into development partnerships, e.g. for hollow floor systems.

Engineer: 'Generally we are involved in projects to specifically do the BIM. In a few cases we are also involved in the further selection of additional parties, such as subcontractors. If we are in the lead we do the selection of the right partners. This is particularly true for our own product and technological development, such as a hollow floor concept.'

3 Integration of operations and processes

An aspect of the BIM approach is that the building parts of all suppliers need to be integrated into the BIM. Standard parts are integrated by the engineer. In case of more complex building parts, the designing and engineering of the supplier of the particular part often has to be part of the designing process of the engineer.

Engineer: 'In the case of the large-scale refurbishment of the monumental building, the supplier of the steel structure was part of the model. This supplier completely designed his steel structure into our model.'

4 Planning and logistics control

Planning and logistics of the project are based on the BIM. When adding time to the model, the planning of the construction of a building is modelled in a 4D model, with time as the fourth D.

Engineer: 'Particularly when we also do the management contracting of a project, we can already start to plan the work in the design and engineering phase of the project.'

5 Quality management

The central issue of BIM is the seamless integration and fitting of all building parts, e.g. via automated clash detection in the BIM model. As a result, quality is implicitly assured by the BIM model. However, after the engineering no quality management along the supply chain has been installed, so this is dependent on the expertise of the contractor.

Engineer: 'Ideally the contractor will follow exactly what has been modelled in the BIM, the virtual prototype of the building. However we are dependent on the contractor and his subcontractors for a good result. We try to influence this process by our integrated planning method and preferably be the management contractor of the work. However, we do not always succeed in this.'

6 Information exchange

All information exchange is connected to the fully integrated BIM model, which includes all designing, detailing and technical data. In addition, the integrated planning method needs to structure information flows along the supply chain.

Engineer: 'If we can apply our BIM and planning method approach fully to a project, in theory nothing should go wrong. In practice, however, misinterpretations or miscommunications often make it harder to actually achieve this.'

7 Product development and design

The design is fully developed and integrated by means of the BIM modelling. This again is an integrated approach per project. In addition, multi-project product development takes place with a floor supplier for a hollow floor concept.

Engineer: 'All our designing is done via BIM. We add engineers and suppliers of building parts to it in case of complex buildings or building parts, but also to calculate the installations or to check connections. Another issue of supplier involvement is standard product development, such as the hollow floor concept we have developed.'

8 Market approach and marketing

Apart from publicity given to the merits of BIM in papers and magazines, and highlighting previous projects and the hollow floor concept, no proactive market approach or marketing takes place.

9 Cultural alignment

No evidence of cultural alignment with supply chain partners was recorded.

Table 27: Case overview of the architectural engineer (Designer 2)

Factors of analysis	Description of application	Reference to 2nd order themes	Development level
Economic governance			
Integration of business activities	In-house construction experts, BIM facilities and in-house management contractor.	Specialist role, absolute control	4. Integrated supply chain
Partner sourcing and collaboration strategies	Partner selection per project. Development partnership with one supplier.	Strategic cooperation	2. Loosely coupled stages
Production management			
Integration of operations and processes	Steel supplier was part of 3D model. No further operational alignment with other partners.	Specialist roles	2. Loosely coupled stages
Planning and logistics control	Planning of projects based on BIM (4D modelling).	Absolute control	2. Loosely coupled stages
Quality management	Quality must be assured by the BIM. No additional quality management along the supply chain.	Real-time information	2. Loosely coupled stages
Inter-firm governance			
Information exchange	Fully integrated BIM including all designing and calculations available.	Real-time information	3. Closely connected stages
Product development and design	Coordination of the design based on BIM. Product development for hollow floor system with supplier.	Design rationality, specialist roles	2. Loosely coupled stages
Market approach and marketing	No proactive market approach or marketing besides publicity for previous projects and hollow floor system.	Technological supremacy	1. Independent stages
Social governance			
Cultural alignment	No cultural alignment with other building partners	Specialist roles	1. Independent stages
Human resource management	No evidence recorded	Specialist roles	1. Independent stages

10 Human resource management

No evidence of joint HR approaches with supply chain partners was recorded.

7.5 Supply chain integration practices applied by builders

7.5.1 Case of a catalogue builder (Builder 1)

In this case, a builder has developed a catalogue approach to house building. The builder has set up a complete factory, including all resources such as concrete fabrication and woodwork. From the factory, the entire process, from sales to transport of ready-made building parts to the site, is organised. The company has built a variety of building types but has now focused on house building, notably individual prefab houses based on the catalogue principle. The builder has integrated the entire process and the supply chain, including a small number of fixed suppliers, with whom long-term relationships have been established (Table 28).

The initial challenge was to build a prefab house in one day after the foundation had been laid. Therefore, a structure was developed whose elements had a maximum size to allow for transport. Once the structure is built, the finishing of the roof and the interior can follow. This approach has resulted in a shorter building time. The focus has been on private clients who desire to build their own house on a plot. The builder has therefore developed a differentiated market approach serving multiple client types, based on age, education, region, family situations and so on. This has enabled the firm to develop corresponding product types per client type, bundled in a catalogue with standard prices, from which clients can choose. The product types and technology (e.g. inner walls) of the houses are flexible to accommodate multiple client wishes.

The builder performs all operations in the company's own factory, integrating the entire process from raw material to the end customer, including the recycling of all waste from the site. The houses are completely prefabricated in the factory, including prefab outer walls with all masonry. The houses are transported in elements at one time to the site, and installed in three days. For an extra fee, the house can be installed in such a way that it can easily be dismantled again. The builder delivers extra options and these, together with the size of the house, determine the price.

The builder has five main suppliers for materials such as bricks and wood. The logistics and planning are optimised exactly to the production at the builder's factory. The relationships with suppliers are strong, and they are regarded as family in a manner of speaking. This creates obligations such as innovations and investments by the suppliers, e.g. technologically and logistically.

1 Integration of business activities

All design, engineering, sales, components manufacture, transport and site construction takes place in-house. This includes the fabrication of concrete including installation, which is rarely done (normally concrete in the Netherlands is acquired from one of the few centralised concrete plants). In addition, the reverse logistics of waste from the site back to the factory is organised by the builder. Back in the factory, the waste is divided and a portion is incinerated to provide heat for the factory.

Builder: 'We do everything ourselves. This is the best way to keep things in control, and that's exactly how we want it.'

2 Partner sourcing and collaboration strategies

Materials for the parts manufacture, such as bricks for the integrated façade elements and wood for the roofs, are sourced from fixed suppliers. All suppliers are contracted for a long period of time and virtually integrated logistically.

Supplier: 'In a manner of speaking, we are regarded as an external business unit for materials supply. Especially for this builder, I am importing certain types of wood I will only sell to him. Often he comes with a particular issue we have to find a technical solution for. So it's much more than just a trading relationship.'

3 Integration of operations and processes

All business and production processes are integrated in-house at the factory location. The process of transport and site construction are also part of the same integrated process and are coordinated centrally and done with the company's own means and people. Suppliers are closely linked to the factory and are treated as external business units.

Builder: 'Since everything and everyone is centrally located or connected to the factory, we can easily control all our operations and make modifications if necessary'.

4 Planning and logistics control

All planning and logistics are centrally coordinated at the factory. The materials supply to the factory is organised in a fixed weekly scheme with weekly orders from the factory straight to the suppliers. The transport of readymade building parts to the site is done with the company's own trucks in a particular fixed scheme to optimally pick up the building parts from the factory and deliver them to site.

Builder: 'We know exactly what to deliver to the site, on what day and in what order. The site construction is strictly organised so we can transport all building parts of a house in 7 to 10 truckloads depending on the house type and the client options, such as with or without a garage'.

5 Quality management

Quality is mainly controlled at the factory when manufacturing the building parts. There are fixed procedures assuring the quality level. The fact that all parts are manufactured in the same factory assures fitting of all parts on site.

Builder: 'Generally when everyone does what he is supposed to do and keeps to the procedures we don't have quality problems. Sometimes something goes wrong but then we exactly know who it was, what went wrong and what to do to prevent it from happening again.'

6 Information exchange

All information is developed sent and controlled from the engineering department of the factory. The centralised control of the information assures all workers in the factory and on site get the right information in time according fixed procedures. This is the same for suppliers.

Builder: 'Everyone gets the same information and thus knows what to do, where and when.'

7 Product development and design

Central to the builder's approach is the catalogue of fully predeveloped products, and the integrated in-house design and production of the houses. This core idea is influencing all other developments within the own business and externally with regards to materials supplied and technological developments.

Table 28: Case overview of the catalogue builder (Builder 1)

Factors of analysis	Description of application	Reference to 2nd order themes	Development level
Economic governance			
Integration of business activities	All design, components manufacture and site construction in house.	Focal firm dominance	4. Integrated supply chain
Partner sourcing and collaboration strategies	All external partners selected and contracted for long time, and virtually integrated.	Strategic cooperation	3. Closely connected stages
Production management			
Integration of operations and processes	All business and production processes integrated in house. Suppliers closely linked to factory. Suppliers are considered 'external business units'.	Operational excellence, absolute control	4. Integrated supply chain
Planning and logistics control	Predefined planning and fully tuned logistics	Inventory control	4. Integrated supply chain
Quality management	Fixed procedures and quality agreements with suppliers assuring quality level.	Total quality	2. Loosely coupled stages
Inter-firm governance			
Information exchange	Information centrally controlled within factory	Absolute control	3. Closely connected stages
Product development and design	Fully predeveloped products in a catalogue and integrated in-house design.	Design rationality, technological supremacy	4. Integrated supply chain
Market approach and marketing	Marketing and sales directly and actively organised from the company's own factory.	Market push, commercial focus	4. Integrated supply chain
Social governance			
Cultural alignment	Suppliers treated as family.	Focal firm dominance, strategic cooperation	3. Closely connected stages
Human resource management	Informal exchange of workforce.	Teamwork	2. Loosely coupled stages

Builder: 'Our catalogue is central to everything we do and develop. All new developments to products or materials and such need to be in accordance with it.'

8 Market approach and marketing

The marketing and sales are fully based on the catalogue approach. All marketing is directly organised from the own sales department towards individual clients. Individual clients are invited and taken through the design and customisation of the own house.

Builder: 'The major part of our sales is to private owners, so our marketing is individualised, in order to service individual owners or families who wish to buy their own house.'

9 Cultural alignment

Employees and suppliers alike are culturally closely bound to the firm, and treated as if they were family. Often events are organised for the people involved and their families.

Builder: 'This is still a family business, and thus must we approach everyone involved including our employees and suppliers as family.'

10 Human resource management

There is no formal exchange of workforce; however, informally the company's own workforce and suppliers do meet in the factory often.

7.5.2 Case of an ICT-driven builder (Builder 2)

In this case a builder has developed his own Building Information Model (BIM) application, with a corresponding process format, and connected partnering suppliers to it. This BIM-based approach to processes and collaboration resolves disconnections of information flows, and supports collaborative working with suppliers. In addition, based on the BIM-system, a corresponding housing concept has been developed with an architect (Table 29).

The system includes a process format supported by 3D BIM for collaboration with suppliers designing online within the 3D BIM model. The investment in the system needs to be recouped by joint benefit from the system on projects. The system must also increase quality and timesaving aspects. The main gain lies in the preparation and shop drawings for a project by modelling the entire house, including all parts and installations. In order to achieve this, the builder has hired an ICT expert to set up the systems and an in-house structural expert. Suppliers were selected who were able and willing to work with the BIM system. The BIM system is able to support the modelling of the structure and a house's installations, but still needs to be made able to model the finishing, based on digital catalogues by suppliers, e.g. kitchen suppliers.

The builder works with product lines with a cooperating architect but is also able to work with other architects. The builder joins the architect in realising his design and advises the client about costs and technology. The builder creates a model based on the architect's design, including all documentation, on a secured website accessible to the suppliers. The builder first puts up the structure of the house and a design protocol. Based on that, the suppliers are able to fill in the design collectively. After this is done, the builder, the architect and the client (e.g. a housing corporation) verify and alter the model as needed. Residents are able to view the design via a 3D simulation viewer. While waiting for the building permit, the builder prepares client options and the purchasing process. Suppliers can prepare to manufacture the parts. At the start of the site construction, the model is fixed, ensuring exact measurements.

Each partnering supplier is responsible for his own part of the work and the coherence of his contribution to the model. The modelling and control are the responsibility of the builder. All the elements of every supplier are added to the digital library: foundation, beams, floors, installations, etc. The result is a blueprint for all projects and all elements that can be used for all types of houses. All partnering suppliers have long-term framework agreements for three years. Price quotations are requested for all projects based on fixed prices, with the aim of lowering them every year by a net reduction of several per cent. Cost reductions are the result of learning and increased productivity. The personnel of the suppliers are trained by the builder to be able to work with BIM. The personnel of the builder and the suppliers working within the BIM environment collaborate and communicate online via web portals and web cams.

1 Integration of business activities

The builder has internalised a fully operational BIM system, ICT capacity, BIM experts and in-house construction engineers. Most notably, the BIM system and experts play a central role in the builder's business process.

BVR: 'Our concept required considerable investment. We applied it first to a part of our business. The rest is still and will probably remain traditional. This also enables us to select the right people for working with the system, while the others work traditionally.'

2 Partner sourcing and collaboration strategies

The system, particularly the BIM part, has required close relations with suppliers, particularly in the technical design and engineering phase. The suppliers are contracted on a long-term basis, educated to work with the BIM model, and required to invest time and effort in order to participate.

Supplier: 'The builder has set an example of working with BIM. For us this was new, and we were willing to step in. This implied more than just a deal; we have been trained by the builder and physically connected with their system. Our man in the engineering is working online with the builder and the other suppliers. Jointly they are engineering and discussing the technical design real-time.'

3 Integration of operations and processes

Besides the online engineering involvement in the technical design via the BIM system, no other processes are integrated with suppliers. The suppliers, however, are motivated to optimise their own production and to preassemble their materials for smooth installation on site.

Supplier: 'What we have been motivated to do, and actually were doing already, is maximising the level of installation of the materials from our factory to site. The BIM system and particularly the exact measurements and knowing the connections with other building parts from other suppliers, has enabled to go further in this than we used to.'

4 Planning and logistics control

Planning and logistics of the materials supply to site are organised based on the BIM system. Connected to the online BIM system there is a shared document server where the planning of the project and further project information are stored.

Builder: 'We gradually want to increase the amount of information shared with building partners, and notably the suppliers. We want the process to become smoother through increased information sharing, and as a result, further improve the preparation and execution of projects'.

5 Quality management

The BIM system must assure better connections between building parts and increase predictability of the execution. As a result, quality must improve. However, no formal, integrated quality management system with the suppliers has been put in place.

Builder: 'Because all suppliers are discussing, engineering and resolving the connections between their respective building parts, the internal connectivity of the design improves. After the engineering, we run a clash detection with a minimum upper limit of a few centimetres or even millimetres, depending on the building parts. This reduces problems with the installation on site, and therefore improves the quality.'

6 Information exchange

Open information exchange with suppliers occurs via the BIM system, including discussion of all designs, calculations and other technical information. Suppliers communicate and cooperate online and real-time via the Internet, mainly via the BIM system, allowing them to take over each other's drawings, share documents via the connected web portal, and communicate directly via web cams.

Supplier: 'It is a great advantage that we can collaborate directly with the others in an open information environment, and adjust and amend information collaboratively.'

7 Product development and design

Besides the BIM system and the collaborative process, the builder has developed a corresponding predesigned house-building concept supported by the BIM system with a partnering architect. The concept is organised in a few product lines of different types of products for different types of users.

Architect: 'The housing concept has been developed in such a way that it optimally connects and makes use of the underlying ICB concept. This has resulted in a few product lines with optional change possibilities for clients such as a housing corporation.'

8 Market approach and marketing

The ICB concept is widely given publicity by the builder. However, the sales approach remains predominantly one-off, although a somewhat repetitive sales approach has been applied by giving the client who did an ICB project a share in the cost reduction, i.e. a price reduction in a following project. The predesigned housing concept is given low profile marketing efforts via a brochure.

Builder: 'The concept is a process concept of being able to handle various projects. With the housing concept, we do not want to become a catalogue builder. However, we would like to achieve a certain level of repetitive sales to the customers we have built for and experience the benefits.'

9 Cultural alignment

Suppliers of the builder are treated as if they were family. Often events are organised with suppliers. In addition, all members of the builder's management team have 'adopted' one of the suppliers and take care of the dialogue and the connection of suppliers to the builder.

Builder: 'We started as a family business. We need to handle our suppliers with care and develop them as if they were family. That's for our and their good.'

10 Human resource management

The joint engineering teams with the suppliers imply a virtual integration of the workforce of the builder and the suppliers. As part of this, the builder also gives combined courses to the personnel of the suppliers, particularly in how to use the BIM system.

Builder: 'Suppliers only had to commit to our new way of working and be willing to adopt BIM. We have taken the lead and taken care of the initial investments, putting in place and coordinating the ICT infrastructure, and delivering courses to the suppliers.'

Table 29: Case overview of the ICT-driven builder (Builder 2)

Factors of analysis	Description of application	Reference to 2nd order themes	Development level
Economic governance			
Integration of business activities	In-house construction expert.	Absolute control	4. Integrated supply chain
Partner sourcing and collaboration strategies	Close relations with suppliers/ co-makers, particularly in design/BIM phase.	Focal firm dominance, strategic cooperation	3. Closely connected stages
Production management			
Integration of operations and processes	Besides design, no other processes integrated.	Operational excellence	3. Closely connected stages
Planning and logistics control	Planning and logistics based on BIM.	Inventory control, real-time information, speed	2. Loosely coupled stages
Quality management	Quality checking via BIM.	Real-time information	2. Loosely coupled stages
Inter-firm governance			
Information exchange	Fully integrated BIM including all designing and calculations. Suppliers participate online via web.	Real-time information, teamwork	3. Closely connected stages
Product development and design	Predeveloped building concept with architect.	Design rationality, market responsiveness	3. Closely connected stages
Market approach and marketing	One-off approach to market. Low profile marketing via brochure.	Market responsiveness	2. Loosely coupled stages
Social governance			
Cultural alignment	Suppliers treated as family. The builder's management have 'adopted' suppliers.	Focal firm dominance, strategic cooperation	3. Closely connected stages
Human resource management	Joint virtual design teams with suppliers. Joint course groups.	Teamwork, specialist roles	3. Closely connected stages

7.5.3 Case of a lean-driven builder (Builder 3)

In this case, the builder has applied a multiple approach to product and process development. The product development includes a low-cost housing concept with an architect. The process development included increasing the level of prefabrication with suppliers, and applying lean techniques to the site construction with subcontractors. For several years the builder has been developing a product and a process concept touching supply chain integration. This includes a predesigned and cost efficient housing concept that is extendable during the life cycle. The process concept is aimed at maximum application of integrated, prefabricated components to increase productivity and site installation speed (Table 30).

In 2004, the builder developed the product concept together with an architect. The primary aim was a cheap housing concept by increasing repetition and strict design organisation. The concept needed to be adjustable to keep residents in their houses as long as possible, even when they wanted to live in a larger house. The concept needed to have many options based on a standard format. In 2006, the concept was put on the market, and in 2009, the builder built a first project with the concept. The concept was developed based on a fixed format but is still flexible externally and internally. Via the Internet, residents are able to assemble their own house, including façade type and style, material use and floor plan, based on budget and taste. The cost of the house is kept low because labour-intensive details and unnecessary cost-increasing elements have been avoided.

The process concept is based on maximum prefabrication to reduce the construction time from 27 weeks to less than 15 weeks, without extra costs. The prefab concrete structure was adjusted to enable the use of larger elements, e.g. three large prefab floor slabs to further increase the speed of construction. Other features, including the prefab bathroom, toilet and meter cupboard, are completely built offsite, sealed and hoisted into the house when the prefab concrete structure is assembled. The toilet and bathroom elements are prepared at the factory to be connected to water and electricity in the house. The inner walls are a ready-made systems wall that can be installed quickly using dry connections. For the electronics, the supplier had developed a prefab meter cupboard including all connections for electricity, telephone, data and media. Wire gutters below the skirting of the walls, and wireless switches for the lighting add to the intelligence of the product and the speed of the process.

1 Integration of business activities

For both concepts, increased coordination and conceptual design was adopted. No activities were internalised. Instead, both architect and suppliers had been given more activities to help in developing both concepts.

Builder: 'We have taken the lead in the development, as it fits well with the lean philosophy we have been following the last few years and want to extend. As a part of this, we need to get closer with our partners and ask them to help us improving the business for mutual benefit.'

2 Partner sourcing and collaboration strategies

Suppliers have been selected and contracted in long-term frameworks. This includes delivering material and equipment, but also implies contributing to the development of the process concept and prefabrication, and associated cost reductions. The architect was involved in the development of the housing concept but has not been involved in the further application of the concept.

Builder: 'We have clearly chosen lean processes and prefabrication of integrated components. This is for direct benefit and useful for existing projects. However, the housing concept is dependent on the market. We have experienced that this may get difficult.'

3 Integration of operations and processes

As part of the lean process approach, subcontractors are an integrated part of the planning and organisation of the construction process on site. Suppliers are also part of the development of integrated components on site.

Builder: 'Although we are in the lead, we heavily rely on our subcontractors and suppliers to get our projects and development activities done. This applies to the day-to-day business, but also innovations in processes and technology.'

4 Planning and logistics control

Planning has been a major issue of the builder's lean management approach. The planning of work on site is based on collaborative lean planning sessions with subcontractors and suppliers. This process is supported by lean planning software that the builder's company had developed specially for them.

Builder: 'Lean is central to our operations, and this particularly applies to the organisation of work on site with subcontractors and suppliers. But to us, lean goes further.'

5 Quality management

As part of the lean approach, the builder has installed a gateway system, meaning that a previous party or person, e.g. a subcontractor, guarantees the quality of the work handed over to the subsequent party or person, i.e. the subcontractor who comes next in the process. Besides those operational measures in the regular project business, the prefabrication, which is part of the process concept, has increased the quality of execution because of the reduction and simplification of site operations.

Supplier: 'The toilets and bathrooms come to the site completely finished off-site, down to the toilet paper holder. The site workers just have to crane it into the structure, connect the water and electricity, and strip the packing foils after finishing, and you're all set. Particularly the bathroom and toilet normally deliver many quality problems, but now this has been resolved.'

6 Information exchange

Besides the shared lean planning, no systematic information sharing takes place in the construction phase with supply chain partners. In the engineering phase, design information is shared mainly with the suppliers, manufacturing the integrated components, e.g. the toilet and bathroom.

Builder: 'We have greatly invested in the lean planning approach, which is the main issue with subcontractors. In the product development we have obviously shared much technical information with the suppliers to develop the prefab components.'

7 Product development and design

Suppliers have been involved heavily in the development of the process concept, notably the development of the integrated prefab building components, i.e. the prefab toilet and bath room, extra-large floor slabs, prefab inner walls and prefab meter cupboard. The architect was also involved and actually led the development of the product concept.

Supplier: 'We have invested quite a bit of time and money in the components. This experiment was one-off. We need a constant stream of work from projects, however, to sustain such developments.'

8 Market approach and marketing

The process development, notably the lean approach, and the prefabrication developments have mainly received attention in professional media. The process concept focuses primarily on housing corporations and their residents and is also marketed via the architect.

Architect: 'Together with the management of the builder we developed the concept. However, I am not sure it will lead to any projects. It depends on the market and the willingness of the builder to push the concept forward whether any projects will emerge. If possible, I would acquire a project myself with the concept.'

9 Cultural alignment

No evidence of cultural alignment with supply chain partners was recorded.

10 Human resource management

The workforce of the subcontractors is part of the joint lean planning and integrated working on site.

Table 30: Case overview of the lean-driven builder (Builder 3)

Factors of analysis	Description of application	Reference to 2nd order themes	Development level
Economic governance			
Integration of business activities	No activities have been internalised.	Focal firm dominance	1. Independent stages
Partner sourcing and collaboration strategies	Co-makers have been selected and contracted in long-term frameworks.	Focal firm dominance, strategic cooperation	2. Loosely coupled stages
Production management			
Integration of operations and processes	Co-makers are an integrated part of the construction process on site.	Operational excellence, teamwork	2. Loosely coupled stages
Planning and logistics control	In particular, the lean planning is a major part of the integrated site management.	Speed	3. Closely connected stages
Quality management	Gateway system in place, i.e. any previous party/person guarantees the quality of the work handed over to the subsequent party/person.	Total quality, absolute control	2. Loosely coupled stages
Inter-firm governance			
Information exchange	Besides the shared planning, no systematic information sharing as yet.	Teamwork	2. Loosely coupled stages
Product development and design	Co-makers and architect were involved in the development of prefab building parts (e.g. toilet, bathroom, large floor slabs, inner walls), and the concept house.	Design rationality, specialist roles, market responsiveness	3. Closely connected stages
Market approach and marketing	Concept house is also marketed via the architect.	Market push	2. Loosely coupled stages
Social governance			
Cultural alignment	No active alignment of organisational cultures	Teamwork	1. Independent stages
Human resource management	Workforce of subcontractors is part of the joint lean planning and integrated working on site.	Teamwork, specialist roles	2. Loosely coupled stages

7.6 Supply chain integration practices applied by suppliers

7.6.1 Case of three suppliers integrating products and site installation (Suppliers)

In this case, three suppliers have integrated their respective components into an integrated housing product, particularly aimed at improved sound insulation and optimised site installation. The lead supplier has also started an in-house subcontractor for the site installation of the sand-lime brick structure of the house as a service to the main contractors (Table 31).

The three suppliers are market leaders for their respective parts. The idea for the integrated housing product was to develop an integrated product specifically to improve the sound insulation of houses. Therefore, heavier types of sand-lime bricks were developed and the connections between the building parts were improved. Another incentive was to improve the competitiveness of sand-lime bricks used for house structures as compared to concrete in larger projects. The integrated product needed to show clients, architects and contractors the benefits of the system in terms of sound insulation, flexibility and price. The aim here was to offer a competitive alternative to concrete for contractors, architects and project developers.

1 Integration of business activities

The three suppliers joined forces to develop the product system, including walls, floors and roof. In addition, they set up an in-house contractor for the site installation, but only for the walls and floors. Besides the delivery and the installation of the product, the engineering department of the lead supplier delivers technical services to architects, developers and contractors to engineer the structure and give advice when and how to apply sand-lime bricks, instead of concrete for instance.

Suppliers: 'We belong to a large supply group. However, the different suppliers are rather separate and not quite used to developing integrated products. The integrated product we developed is one of the first attempts. The installation service is a separate path and not connected to the integrated product as yet because of different technologies applied. The integration of the site installation implies taking this over from the contractor, however, as a service to contractors.'

2 Partner sourcing and collaboration strategies

In the case of the integrated product, the suppliers are part of the same supply group and thus automatically linked. However, no extra efforts were made for intensified collaboration. The installation service is part of the lead supplier, often collaborating with the other suppliers but also with external suppliers and specialist subcontractors, e.g. for gluing the sand-lime bricks. Those specialists are contracted based on yearly contracts.

Suppliers: 'One would expect that we would automatically source from other suppliers in the group. However, this is not always the case. Although formal links exist automatically, group companies need to make as much effort connecting with other group companies as with external firms, which are often competitors. This has never been a problem for us or for our competitors.'

3 Integration of operations and processes

The processes of the three suppliers are separate, except for the engineering, which is centralised at the lead supplier. The integrated product and the installation service have not been integrated as yet because of technical impossibilities resulting from the application of different floor types in both systems. The product system is aimed at sound insulation, so it has increased the weight of the structure, including the floor. The installation service, on the other hand, has to use lighter floor types. Part of the installation service installation operations on site are integrated, which is not the case for the product system.

Suppliers: 'The aim and the approach of the product system and the installation service are different. Therefore, both systems can't be integrated as yet.'

Contractor A: 'The product is technically okay, but the fragmented approach by the three suppliers to organising the logistics was not satisfying. They should integrate those aspects, too, and preferably install the entire product on site for contractors. That would help.'

4 Planning and logistics control

The operational planning is coordinated by the lead supplier on behalf of the three firms. However, the logistics and transport of the parts from the three firms to the site are separate and need to be coordinated by the contractor. In the case of the installation service, the planning of the site installation is coordinated with the contractor. However, in this case as well, the delivery of the sand-lime bricks and the floors are coordinated by the respective suppliers of those parts separately with the builder.

Suppliers: 'We haven't resolved the delivery issue to the site yet. We understand this could be more centralised and coordinated jointly by the suppliers. This reflects the relative connection or disconnection between supply firms of the group.'

Contractor B: 'The process went quite smoothly. We didn't have to care about the installation of the structure. The whole installation was done by the supplier without any interference on our part, other than the planning and preventing logistical clashes of the installation related to the other activities on site.'

5 Quality management

No joint quality management has been put in place for the integrated product. However, the central engineering by the lead supplier secures the quality of the integrated product. In the case of the installation service as well, the quality is controlled by the lead supplier. Therefore, the lead supplier connects or takes part in the quality system of the contractor. The installation service has its own staff for project control and registration of complaints. Based on the complaints, quality problems are resolved together with the contractor, whether they pertain to the product or the process.

Suppliers: 'We would often get complaints about the quality because of installation errors by the contractor. Now with the integrated approach and with us taking care of the installation, this has become much better.'

6 Information exchange

Besides the centralised engineering of the integrated product, no intensified information sharing takes place between the three suppliers. In the case of the installation service, the information exchange with the contractor is quitc intensive. Shop floor drawings are sent back and forth to the contractor. In a few cases and as an extra service, the lead supplier took care of all the drawings on behalf of the contractor. Coordination of drawings takes place digitally. Specialist subcontractors for the site installation are directly connected to the information.

Suppliers: 'We have experienced improved communication with contractors. This gave us better results. However, this could still be much improved.'

7 Product development and design

The integrated product is a joint development, and the design is integrated by the lead supplier. The installation service covers only the walls and floors. Ideally, the service would include the façades and roofs in order to deliver and install a complete and finished product.

Suppliers: 'Our products are technically of a high standard. However we need to develop more complete and integrated products for our customers.'

8 Market approach and marketing

The integrated product is marketed as such to builders and project developers. The three suppliers have tried to respond to this trend. The lead supplier aims in the near future to develop complete pre-glued wall elements of sand-lime bricks to be able to compete better with prefabricated concrete.

Suppliers: 'We keep developing our products and communicating this with the market. We have experienced that sand-lime bricks in the eyes of many of our clients is believed to be more expensive and less practical than concrete. We need to overcome that.'

9 Cultural alignment

Within the supply group the three suppliers belong to the same corporate culture; however, this is a formality, less obvious in practice. In the case of the installation service, specialist subcontractors are connected and treated as external business units. On the client's part, the relationship is aimed at delivering service.

Suppliers: 'We want to share the same values as our clients, and we also want our partners to share our values.'

10 Human resource management

In the case of the installation service, personnel on site join into combined installation crews of the personnel of the lead supplier, the specialist subcontractors, and the contractor, i.e. the client.

Table 31: Case overview of the three suppliers integrating products and site installation (Suppliers)

Factors of analysis	Description of application	Reference to 2nd order themes	Development level
Economic governance			
Integration of business activities	Three subsidiaries of a supply group jointly developed an integrated product (walls-floors-roof), and set up an in-house installation service for site installation of walls and floors for contractors.	Strategic cooperation, market responsiveness	4. Integrated supply chain
Partner sourcing and collaboration strategies	As part of the supply group, the firms are automatically linked. However, no extra effort for intensified collaboration is made.	Strategic cooperation	3. Closely connected stages
Production management			
Integration of operations and processes	The processes of the firms are separate, besides the design, which is centralised at the lead supplier.	Specialist roles	2. Loosely coupled stages
Planning and logistics control	Planning is integrated by the lead firm with contractors, but the logistics to site are separate.	Inventory control	2. Loosely coupled stages
Quality management	In the case of the installation service, quality is controlled by the lead supplier with the contractor on site.	Total quality	1. Independent stages
Inter-firm governance			
Information exchange	Besides the joint design there is no intensified information sharing.	Teamwork	1. Independent stages
Product development and design	The integrated product is a joint development, and the design is integrated by the lead firm.	Design rationality, specialist roles	3. Closely connected stages
Market approach and marketing	The product is marketed as an integrated, technologically innovative solution, however passively.	Technological supremacy	2. Loosely coupled stages
Social governance			
Cultural alignment	As part of the installation service, specialist subcontractors are treated as external business units.	Focal firm dominance, Strategic cooperation	2. Loosely coupled stages
Human resource management	Joint installation crews on site of the personnel of the lead supplier, specialist subcontractors and contractor/client.	Teamwork, specialist roles	2. Loosely coupled stages

7.7 Reflection on key issues and contributions of this chapter to the thesis

The foregoing descriptive exploration of the construction cases is summarised in a brief overview of applications of supply chain integration as found to be applied by the five different types of firms observed (Table 32). For each case, the focus, extent and level of the issues addressed vary considerably. This variance among the cases could be clarified based on the different businesses, positions and capabilities of the respective firms observed in the cases, in relation to the other firms in the supply chain in which they are active.

The representation of supply chain integration in the building industry must be approached in a differentiated manner depending on the type of focal firm. This is different from the manufacturing cases where the lead manufacturer is normally the party acting as a focal firm. These characteristics and differences, and the effect they have on the various representations of supply chain integration in the building cases versus the manufacturing comparators, give initial direction to the further analysis of the empirical findings and the synthesis of the theoretical contribution in the following part of the thesis.

Table 32: Summary of applications of supply chain integration found in the building cases

Factors of analysis	Clients	Project developer	Designers	Builders	Suppliers
Economic governance					
Integration of business activities	• Reduction of activities, shifting to contractors • Legal partnership with municipality and contractors	• Integration of life cycle • Integrated delivery of facility services with the built asset	• Integration of developer, engineer, management contractor in house • In-house BIM system and expertise	• Sales, design, engineering, parts manufacture, transport, site installation in house • In-house BIM system and expertise	• Three suppliers delivering one product • Supplier also installing product on site
Partner sourcing and collaboration strategies	• Long-term relations with contractors • Aimed at cost reductions	• Fixed relations when possible	• Project partners selected per project • Product development with suppliers	• Suppliers contracted long term • Close collaboration with suppliers	• Suppliers part of supply conglomerate • Fixed subs for installation of product
Production management					
Integration of operations and processes	• Joint overall development of projects with contractors	• Integrated control of entire real estate process	• Loosely operational links	• Integration of all operations in factory • Operational integration on site via lean planning with subcontractors	• Centralised engineering of the product • Site installation jointly with subs
Planning and logistics control	• Planning and logistics by contractors • Lean approach to site philosophy as a pointer for builders	• Overall planning in house • Management contracting in house • Site logistics by contractor	• Planning based on BIM • Management contracting in house	• Planning and logistics from the builder's own factory, own transport • Planning based on BIM • Lean planning on site • Transporting and installing large integrated components on site	• Planning jointly with main contractor
Quality management	• No explicit supply chain approach	• No explicit supply chain approach	• Quality assured via integrated BIM	• Total quality management in factory • Quality assured via integrated BIM • Gateway quality system on site	• No explicit supply chain approach

Inter-firm governance					
Information exchange	• No explicit supply chain approach	• No explicit supply chain approach	• Information exchange via BIM	• Close information in factory, fixed suppliers • Information exchange via BIM	• No explicit supply chain approach
Product development and design	• Project development jointly with contractors • Architect not involved in partnership	• Product development in house • External architect for design	• Product development with specialist suppliers • Design and engineering in house	• Catalogue approach of predeveloped designs • Predeveloped housing concept with architect • Advanced prefabrication with suppliers	• Joint product development of one product • Taking over parts of the engineering from the contractor
Market approach and marketing	• Residents involved in planning of refurbishment	• Being your own client • Delivering full service to user	• One-off projects • Few framework agreements for multiple projects	• Marketing and sales directly from factory • Repetitive sales via price reductions • Marketing also via architect	• Marketed as integrated and technologically innovative product
Social governance					
Cultural alignment	• Contractors, subs, residents involved in movement	• No explicit supply chain approach	• No explicit supply chain approach	• Suppliers regarded as family • Management adopting suppliers	• No explicit supply chain approach
Human resource management	• Joint staffing on management board and on projects	• No explicit supply chain approach	• No explicit supply chain approach	• Interaction with suppliers • Joint virtual engineering via Internet	• Joint workforce on site with subs

PART III:

SYNTHESIS

8 Case study analyses

In this chapter, the empirical data gathered from the comparator studies in manufacturing and the case studies in the building industry presented in Chapters 6 and 7 are analysed and compared. Following Eisenhardt (1989), extensive narratives and iterative tabulation will be used to discern patterns across the findings. First, the building case studies and the manufacturing comparator studies will be analysed separately. Further insight from the building case studies and the manufacturing comparator studies is generated by within-case and cross-case analyses. Next, the aggregate findings from the building case studies are compared to the aggregate findings from the manufacturing comparator studies. The aim is to explain and understand the differences between supply chain integration applications in the building cases versus manufacturing. This gives initial directions for the development of the concept of supply chain integration for the building industry, i.e. shaping the hypotheses in the following chapter.

8.1 Analyses of the building case studies

8.1.1 Within-case analyses of the building cases

The within-case analysis is aimed at mastering the data gathered in the case studies, and further understanding the unique features of each of the individual cases (Eisenhardt 1989).

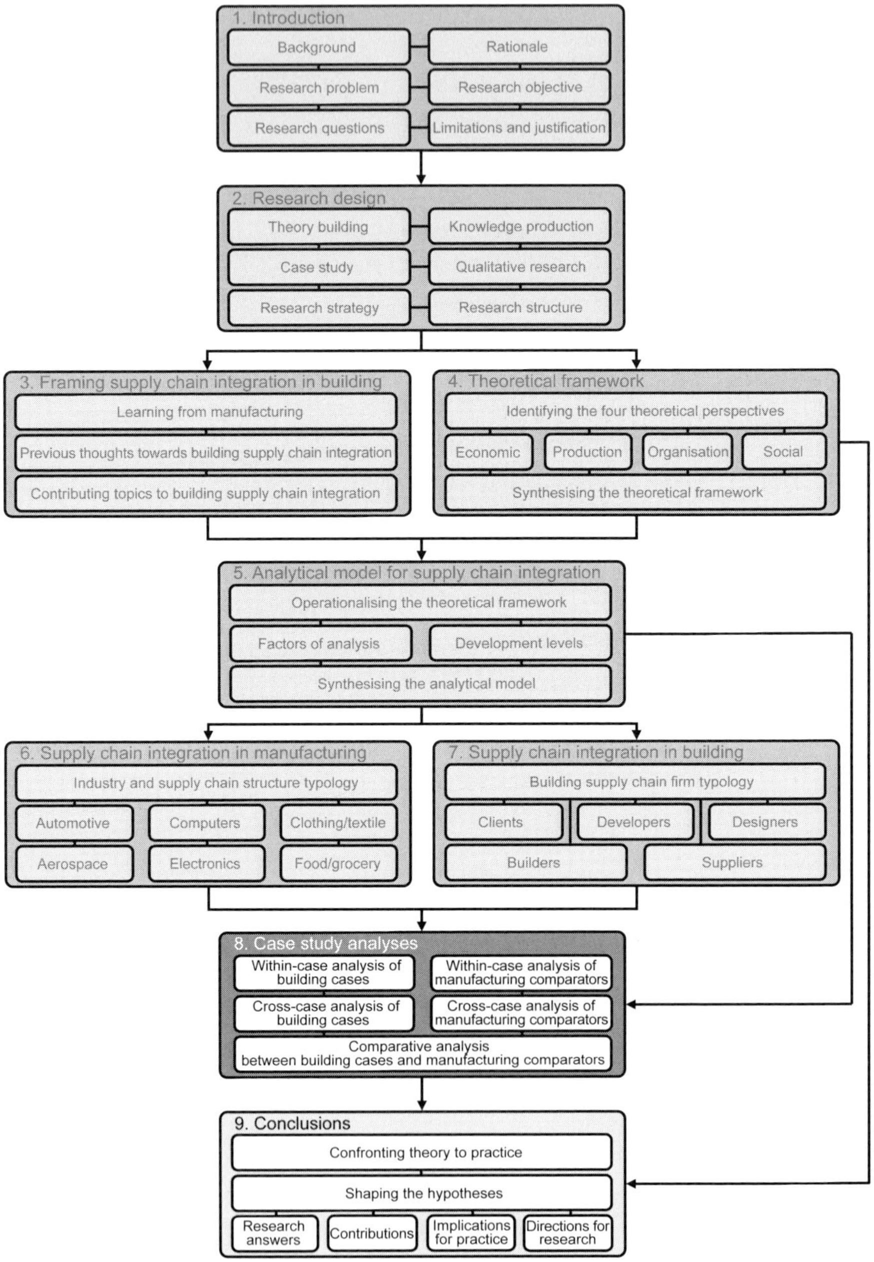
1. Introduction
Background
Rationale
Research problem
Research objective
Research questions
Limitations and justification
2. Research design
Theory building
Knowledge production
Case study
Qualitative research
Research strategy
Research structure
3. Framing supply chain integration in building
Learning from manufacturing
Previous thoughts towards building supply chain integration
Contributing topics to building supply chain integration
4. Theoretical framework
Identifying the four theoretical perspectives
Economic
Production
Organisation
Social
Synthesising the theoretical framework
5. Analytical model for supply chain integration
Operationalising the theoretical framework
Factors of analysis
Development levels
Synthesising the analytical model
6. Supply chain integration in manufacturing
Industry and supply chain structure typology
Automotive
Computers
Clothing/textile
Aerospace
Electronics
Food/grocery
7. Supply chain integration in building
Building supply chain firm typology
Clients
Developers
Designers
Builders
Suppliers
8. Case study analyses
Within-case analysis of building cases
Within-case analysis of manufacturing comparators
Cross-case analysis of building cases
Cross-case analysis of manufacturing comparators
Comparative analysis between building cases and manufacturing comparators
9. Conclusions
Confronting theory to practice
Shaping the hypotheses
Research answers
Contributions
Implications for practice
Directions for research

The analysis of each case study is based on a *descriptive-analytical approach*. This implies a discussion of the case studies, and particularly finding *relations between actions and their effects* as identified in the case studies. The actions are related to the factors of supply chain integration. The effects havc been categorised according to four aspects: effects on the delivery to clients, effects on the building process as a whole, effects on internal processes and productivity, and financial effects.

Case of the housing corporation in operational cooperation with builders (Client 1)

The housing corporation acted as the focal firm promoting integration and repetition among supply chains of main contractors, specialist contractors and suppliers (Table 33). The suppliers had to integrate the supply chain themselves and qualified for long-term contracts for multiple projects. The supply chain firms participated actively in the planning of projects and work, and were involved collaboratively in the price setting and contracting of the projects. The clients of the housing corporation were actively involved in the definition and planning of the projects, and interacted with and judged the supply chain firms.

Table 33: Within-case analysis of the case of the housing corporation in operational cooperation with builders (Client 1)

Aspects	Actions	Effects
Product delivery and client satisfaction	• Framework agreements had been established with four maintenance firms for the maintenance of houses, and initially nine contractors and thirty specialist subcontractors for refurbishments. • Residents have been intensively involved in the planning and the execution of projects. After the completion of a project, residents were involved in the evaluation of the particular contractor of the project.	• Budgets for the client organisation were secured. • Residents were satisfied with their involvement in the planning of the work.
Building process	• Projects were left to the contractors to manage. Planning and logistics were based on a lean construction approach by the contractors and subcontractors collaboratively. • As a result, site logistics as well as supply logistics were closely discussed by the contractor with subcontractors and suppliers.	• Effects on the building process included smoother and faster processes.
Internal processes and production output	• Reduced own effort by involving contractors in a collaborative procurement setting.	• More effective internal processes and production resulted in faster work on site for contractors. • Increased productivity and reduced capacity than normally needed to coordinate projects by the client.
Financial effects	• Frequent review meetings were aimed at the establishment of an open and informal relation between representatives of the client organisation and the contractors.	• Financial effects included less price fluctuation, fewer claims for extra work, and higher profitability. • Improved exposure and image for the firms involved, and improved client satisfaction.

Case of the housing corporation in legal partnership with builders (Client 2)

The housing corporation joined into a public-private partnership with the municipality and with the contractors (Table 34). The partnership was aimed at financially beneficial planning of multiple projects for a long period of time. The supply chain integration remained on a strategic level among those parties and did not touch operational issues, nor did it involve additional firms, such as architects or subcontractors. The partnership did not aim at integration of the delivery process towards clients of the housing corporation.

Table 34: Within-case analysis of the case of the housing corporation in legal partnership with builders (Client 2)

Aspects	Actions	Effects
Product delivery and client satisfaction	• The overall planning and the procurement of the projects take place in a joint effort with the municipality, the corporation and the contractors. • The aim is to achieve budget-neutral tracks of multiple projects.	• Project developments are secured, prices are stable and profits are low, due to the financial balancing between the projects into the budget-neutral tracks.
Building process	• The corporation initiated its supply chain integration approach particularly on an institutional level together with the municipality. • Multiple projects were planned within the framework of a 15-year Public Private Partnership with the four contractors involved.	• Effects on the building process are minimal, and the building process is still somewhat traditional.
Internal processes and production output	• Based on the overall planning, the projects are developed collaboratively per individual project.	• There are no effects on internal processes and production output, because there is little attention paid to gains such as improvement of productivity.
Financial effects	• Builders involved accepted a lower profit margin based on the long-term perspective and the continuity of work.	• Financial effects included the relatively low project budgets for the client organisation, because of the lower profits and the budget neutralisation between projects, and greater continuity for the contractors.

Case of the full-service project developer (Developer)

The developer acted as the focal firm, following an internalisation approach leading to integrated project delivery, including facility management, to the client (Table 35). In addition to facility management, the integration included energy delivery to buildings as well as additional services delivered by the developer to the user, aimed at a reduction of the total cost of ownership and an increase of the life cycle service level. On the other hand, the centrally controlled design and build processes, and the corresponding involvement and contracting of supply chain firms remained rather traditional.

Table 35: Within-case analysis of the case of the full-service project developer (Developer)

Aspects	Actions	Effects
Product delivery and client satisfaction	• Internalisation, and fixed relationships are sought where possible with architects, builders and energy suppliers, e.g. for energy delivery with the built assets they deliver. • The project developer coordinates all processes linked to the real estate process, including the design and build done by external partners.	• Effects on product delivery and client satisfaction are reflected in the full-service approach to clients who are offered an integrated product, including all facilities and services.
Building process	• The general project planning is coordinated by the project developer, and the planning of the work and logistics on site are coordinated by the contractor of the project.	• Effects on the building process are limited, and the building process is still somewhat traditional.
Internal processes and production output	• The product development and concept design take place in house and proactively, well before a project is in sight. • The products and concepts developed are leading for new project developments. Within the project development, the architecture and internal design of buildings are outsourced to contracted designers.	• There are no effects on internal processes and production output as no emphasis is put on those issues.
Financial effects	• The project developer aims for integration or at least dominant coordination of the life cycle on all its projects from land acquisition to facility management, including all intermediate phases. • Preferably, the project developer is its own client organisation, and from that position delivers the building, including the total facility management, to the end user.	• The financial effects of the integrated approach are generally decreased price levels because of package deals for a building, including facilities management, which in turn generates extra return.

Case of the developing and contracting architect (Designer 1)

The architect acts as the focal firm, integrating project development, design, engineering and management contracting (Table 36). This enables the firm in particular to effectuate and control its own designs from the early stages of a project until the construction on site. For the manufacture of parts and site assembly, the firm controls the process, aiming at speed and efficiency, improved buildability of the design, and maximum prefabrication. The supply chain firms are contracted in principle on a project basis, but apparently, this is mainly a fixed pool of firms.

Table 36: Within-case analysis of the case of the developing and contracting architect (Designer 1)

Aspects	Actions	Effects
Product delivery and client satisfaction	• The project planning and management are coordinated from the design, and are passed on via the in-house contractor to the external contractor who actually builds the project. • The in-house management contractor coordinates up to the site logistics.	• Clients can procure the entire track, including project development, design, build and supply.
Building process	• Selection of other building partners in principle takes place per project; however, relations have sprung up with informal preferred partners. • Parts development, however, does take place in strategic partnership with specialist suppliers, e.g. collaborative development of an integrated climate floor system.	• Effects on the building process imply a building process that is much more driven by architecture and product development than by operational issues.
Internal processes and production output	• The architecture firm has internalised a project developer and a management contractor within its own organisation. • Within the architecture firm, the developer, the designer and the management contractor are separate units, thus no operational links exist among the three.	• Effects on internal processes and production output include increased speed of building and productivity on site because of the application of modular design, extensive prefabrication, and fast installation on site.
Financial effects	• Most projects are approached as one-off, some as a package, e.g. the building of multiple logistics centres for a postal services firm.	• Financial effects, particularly in the case of the repetitive projects, include higher profitability because of the integrated approach and repetitive use of one design blueprint for multiple projects.

Case of the architectural engineer (Designer 2)

The engineer has integrated architectural services internally and can also act as developer and management contractor (Table 37). The integration approach is based on Building Information Modelling (BIM). The delivery process is integrated and controlled by the BIM model. Supply chain firms are connected to the BIM model. Their involvement in the design and engineering ensures buildability. The firm controls the engineering and the presentation of the design to the client and the user. The firm has not established strategic collaborations with supply chain firms.

Table 37: Within-case analysis of the case of the architectural engineer (Designer 2)

Aspects	Actions	Effects
Product delivery and client satisfaction	• The planning of projects is based on the 3D BIM model with time added as a factor, i.e. 4D modelling. • An integrated quality approach has been put in place in such a way that the integrated design coordination by the 3D BIM model assures the quality and consistency of the design.	• With regard to effects on product delivery and client satisfaction, clients are informed early about visuals and technical feasibility of a project by the 3D BIM model.
Building process	• An initiative for joint, long-term product development has taken place in partnership with a supplier of a hollow floor concept. However, most collaboration with firms takes place on a project basis. • In some cases, suppliers have been involved quite intensively in the engineering via a 3D BIM model, for instance, the supplier of a complex steel structure.	• Effects on the building process include the improved buildability of designs and increased possibilities for prefabricating building components and more complex building parts, and their exact fitting on site.
Internal processes and production output	• The firm has internalised in-house structural experts, BIM experts and ICT facilities, as well as an in-house management contractor. • The core of activity is BIM-supported architectural engineering. • The contracting has remained modest.	• Effects on internal processes and production output mainly pertain to prevention of the need to rework the design on site.
Financial effects	• The model integrates HVAC calculations and engineering of installation.	• Improved effectiveness and efficiency of work and the resulting workload reduction have led to reduced operational costs.

Case of the catalogue builder (Builder 1)

The builder has internalised the entire manufacture of prefabricated building parts and the design and building of predesigned houses, applying industrialised building, standardised parts and off-site production (Table 38). Production takes place in a centralised factory. Strategic collaboration has been established with suppliers who deliver directly to the factory. Transport to the site and assembly is fully harmonised with the production. The entire process is centrally controlled. The catalogue approach with standard prices enables structured client contact.

Table 38: Within-case analysis of the case of the catalogue builder (Builder 1)

Aspects	Actions	Effects
Product delivery and client satisfaction	• The factory is an integrated facility where orders from customers on the one side and supplies from suppliers on the other are directly taken in and processed into a fully equipped house. • The market strategy is applying a catalogue approach of fully predesigned and technically predefined products. • All marketing and sales are organised directly from the factory.	• The effects of the approach on product delivery and client satisfaction are such that clients are absolutely sure about design and price, and a smooth product delivery is guaranteed.
Building process	• Suppliers are contracted for long periods of time, virtually integrated as external business units, closely linked to the factory and treated as family. • The planning, production, logistics and transport are fully coordinated from the factory. • Transport from the factory to the site is organised efficiently in combined shipments.	• Effects on the building process are mainly concentrated in highly optimised effectiveness and efficiency while control of the entire building process is in one company's hands.
Internal processes and production output	• The builder has internalised all sales, design, engineering, parts manufacture and site construction in house in a factory environment. • Within the factory, the conditioned production environment and fixed procedures assure the quality level of the production. • Information is centrally controlled at the factory office and sent directly to the shop floor.	• The effects on internal processes and production output include high productivity in the factory as well as on site.
Financial effects	• Resources and activities are internalised to the maximum. • Logistics and reverse logistics from the factory to the construction sites are organised in house. • Waste is incinerated in the factory to heat the factory building with virtually no use of external resources such as natural gas.	• The financial effects include higher profitability and cost efficient building because of the repetition of products via the catalogue approach. • There are positive effects on sustainability because of the recycling of waste and the efficient organisation of transport.

Case of the ICT-driven builder (Builder 2)

The builder as a focal firm controls the engineering and ideally the design, applying BIM to control the site construction and coordinate the parts manufacture collaboratively with suppliers (Table 39). Working jointly with suppliers has further been supported by means of ICT. For the design, the firm has established a strategic collaboration with an architect developing predesigned yet customisable housing products. The client shares in cost reductions achieved during the project, as do the suppliers.

Table 39: Within-case analysis of the case of the ICT-driven builder (Builder 2)

Aspects	Actions	Effects
Product delivery and client satisfaction	• The BIM model assures the quality and consistency of the design and supports the quality of the work.	• With regard to the effects on product delivery and client satisfaction, fast and fully engineered preparation of projects reduces the lead-time of projects.
Building process	• Establishment of close relations with suppliers, particularly in the engineering. • Suppliers are involved on-line and real-time in a joint engineering environment by the builder, and all are working within the same BIM model. • All parts manufacture and the planning and logistics are based on the BIM model. • Close relations have been established with an architect with whom a predesigned building concept has been developed, connected to the BIM system.	• Effects on the building process pertain to the high levels of interaction with suppliers in the engineering phase, and the resulting improvement of efficiency and speed of operations.
Internal processes and production output	• The builder has internalised an in-house structural engineer, BIM experts, BIM systems, and ICT infrastructure.	• Effects on internal processes and production output include higher productivity during the engineering and on site.
Financial effects	• Repetitive projects have been applied to achieve cost and price reductions for supply chain partners as well as the client on a subsequent project based on the cost reduction achieved in the previous project.	• Financial effects include cost reductions and increased profitability.

Case of the lean-driven builder (Builder 3)

The builder acts as a focal firm, applying lean techniques with suppliers and productivity improvements on site via process and product innovations, including prefabrication of integrated parts and components (Table 40). The parts and components were fully developed, engineered and manufactured in close collaboration with fixed suppliers. In this case as well, the firm has established strategic collaboration with an architect for the development of predesigned yet customisable housing products.

Table 40: Within-case analysis of the case of the lean-driven builder (Builder 3)

Aspects	Actions	Effects
Product delivery and client satisfaction	• A predesigned housing concept had been developed and marketed with an architect for newly built housing. • A pre-engineered modular concept for refurbishments had also been developed with suppliers, to allow prefabrication of complete interiors in large, integrated modules.	• This approach has led to increased certainty about price, design and construction time for clients.
Building process	• Suppliers and subcontractors have been selected and contracted in long-term framework agreements. • Subcontractors are an integrated part of the construction process on site. • The lean planning and coordination on site are a major part of the integrated site management. • The firm has developed its own planning software to support the lean planning approach.	• The effects on the building process have implied high levels of involvement of subcontractors as part of the lean approach, resulting in increased productivity and efficiency.
Internal processes and production output	• A quality system has been put in place, implying that previous subcontractors guarantee the quality of the work delivered and handed over to subsequent subcontractors.	• Effects on internal processes and production output include faster and qualitatively improved site work.
Financial effects	• Suppliers have been involved intensively in the development of a fully prefabricated modular approach of integrated building parts, such as completely finished bathroom and toilet modules.	• Increased quality and reduction of reworking and work on site have resulted in cost reductions.

Case of the three suppliers integrating products and site installation (Suppliers)

The three suppliers have integrated the engineering and manufacture of their three parts – floors, walls, and roof – resulting in an integrated product coordinated by the lead supplier (Table 41). The integrated product is particularly aimed at contractors. In addition, site installation is a service the suppliers deliver to the contractor. The installation is handled by a specialised installation firm. The suppliers and the installation firm have been involved for a long term. Collaboration with contractors has often also been intensive and repetitive.

Table 41: Within-case analysis of the case of the three suppliers integrating products and site installation (Suppliers)

Aspects	Actions	Effects
Product delivery and client satisfaction	• The installation firm connects to the contractor of a project, and takes care of the complete installation on site with the firm's own installation crew.	• The integrated engineering and installation of a complete structure has led to a reduction of coordination needed by the contractor.
Building process	• The three suppliers have developed an integrated product for the complete structure of a house.	• Effects on the building process are an increased integration of suppliers and contractor, and a shifting of activities from the contractor to the suppliers.
Internal processes and production output	• The processes of the supply firms have remained separate, apart from the centralised engineering of the integrated product. • The planning is also centralised; however, the logistics of the materials shipped to the site is separate.	• The effect on internal processes and production output pertains to increased productivity in the site installation.
Financial effects	• The in-house contractor pools the company's own resources with those of subcontractors as if they were external business units.	• The financial effects include cost reductions of the site installation.

8.1.2 Cross-case analysis between the building cases

The cross-case analysis is aimed at identifying similarities and differences between the groups of building cases, and matching patterns between them (Eisenhardt 1989).

Similarities between supply chain integration practices applied in the building cases

In most cases, the supply chain approaches of the different firm types have not included an integrated solution for *quality management* nor for *information management*. Additional aspects which were generally missing are *cultural alignment* or integrated approaches to *personnel* among firms in the supply chain. In contrast, the focal firms have in most cases applied a certain kind of *internalisation* of activities, whether into their own organisations or via a joint legal entity with supply chain partners. Besides the tendency to internalisation, in all cases the focal firms have been looking for long-term *collaborative arrangements* with firms in the supply chain. Further similarities could not clearly be found between the approaches of the different firm types.

Differences between supply chain integration practices applied in the building cases

The differences between the case studies mainly reflect the different approaches to supply chain integration between the various firm types. Clearly, the *client organisations* observed have tried to shift activities towards contractors via fixed arrangements mainly aimed at cost reductions. The *project developer* aimed at an integrated approach to delivering the entire product, including all services during the use and facility management. The *designers* concentrated on coordinating all processes around the design towards development as well as engineering and contracting. The *contractors'* supply chain approaches aimed at increased levels of predevelopment of conceptual products, and the full coordination of site construction and supply. The *suppliers* focused on the delivery of pre-integrated products from different parts, including the delivery of engineering and site installation services.

In addition, many differences could be found between the approaches of the different firm types (Table 42). The clients tended to *dominate* while cooperating strategically, primarily with builders, and trying to establish a higher level of *teamwork* with them. The developer in particular stressed a *dominant approach* in pushing the market, combined with a highly responsive and *commercial focus* on its business and a desire to be in *complete control* of that business. The designers focused on their *specialist role*, pursuing high *technological* standards, seeking rationality in their *designs* and engineering the designs with suppliers. The builders have collectively shown evidence of addressing virtually all issues, *particularly towards suppliers*, except for taking on a specialist role, which is understandable from the viewpoint of a main contractor who generally outsources this role to subcontractors and suppliers. The

Table 42: Similarities and differences between the findings from the building case studies based on 2nd order themes

Themes	Clients	Project developer	Designers	Builders	Suppliers
Focal firm dominance	Corporation leading builders	Developer controlling supply chain	Designers did not dominate	Builder integrating supply chain	Suppliers did not dominate
Market responsiveness	Users involved in the project definition, planning	Developer anticipating client	Designers followed market	Builder offering products proactively	Suppliers anticipating contractors
Operational excellence	Clients not involved in operations	Developers not involved in operations	Designers not involved in operations	Builder internalising operations	Operational improvements no issue
Market push	No proactive market strategy	Developer acting as client delivering full service	Designers followed market	Builder launching products	Suppliers followed market
Commercial focus	No commercial activities	Developer actively marketing	No commercial activities	Builder actively marketing	No commercial activities
Speed	Speed as a target for contractors	Speed no issue	Speed no issue	Speed as an operational goal	Speed minor issue
Absolute control	Leaving control to contractors	Developer exercising commercial power	Design strictly controlled, but not operations	Builder controlling entire building process	No power to exercise control
Real-time information	Information not managed as a key issue	Information not managed as a key issue	Designer integrating information directly in BIM	Builder steering information directly to shop floor	Dependent on information from contractors
Inventory control	Inventory control no issue	Inventory control no issue	Inventory control no issue	Builder controlling stocks with suppliers	Suppliers controlling stocks with builders
Specialist roles	General role played, leaving this to contractor	General role played, leaving this to contractor	Designer pursuing BIM specialisation	Subcontracting to specialist subcontractors	Suppliers adding technological expertise
Total quality	Leaving quality to contractors	Commercial quality important, leaving technical quality to contractors	Design quality important, leaving technical quality to contractors	Builder managing quality with supply suppliers	Suppliers following contractors and own quality system
Design rationality	Design rationality no issue	Design rationality no issue	Designers pursuing rational design and process	Builder pushing product development and prefab	Suppliers pushing technological and operational rationality
Strategic cooperation	Corporation cooperating with builders	Predominantly focusing on internalisation	Joint product development with specialist suppliers	Builder strategically sourcing suppliers	Suppliers entering in strategic collaboration
Technological supremacy	Technology no issue	Technology no issue	Designer pursuing high-tech design, material use	Builder pushing product and process innovations	Suppliers improving material technology
Teamwork	Joint staff with builders in management	Predominantly focusing on own personnel	Incidental teamwork with specialist suppliers	Collaborative engineering with suppliers	Joint workforce on site with contractor and subcontractors

suppliers were found to concentrate mainly on their respective *specialist roles* within a mutually collaborative setting, and on addressing issues within their particular domain of materials *technology*, rational components design and *inventory control*.

The *development levels* per case have been assessed based on the observation of the various supply chain integration practices in the different cases (Table 43). However, the manifestation of those practices varies quite a bit from case to case. The assessment of the levels of the different cases therefore is indicative and not absolutely comparable. Although the cases could be on the same level for certain factors, the particular applications will still differ.

Apparently, the average of the building cases mainly conforms to the second level of 'loosely coupled stages'. However, one factor deviates from this pattern. The high score of many cases for the factor 'Integration of business activities' can be explained by the general preference of focal firms to pursue internalisation of business activities into their own organisation. Alternatively, there is a tendency among firms to choose to *integrate joint business activities* in a separate organisational unit or a joint legal entity. Such solutions have traditionally often been the case temporarily in the building industry for large project organisations. Apparently, firms tend to apply this solution also for joint undertakings in the light of supply chain integration. In addition, a case that scores high for most factors of analysis is the case of the *catalogue builder*, since the particular firm has factually integrated most business and production activities in house. Obviously, this application can generally be regarded as a highly advanced but rare example of supply chain integration in the building industry.

8.2 Analyses of the manufacturing comparator studies

8.2.1 Within-case analyses of the manufacturing comparators

Like the building case studies, the manufacturing comparators are analysed separately as well as comparatively, in order *to be able to draw comparisons* between both building and manufacturing. To this end, the manufacturing comparators collectively function as an 'aggregate comparator' to which to compare the building case studies in the following section of this chapter.

Automotive industry, notably Toyota

Car manufacturers like Toyota have particularly aimed their supply chain strategies at their supply base. Suppliers are fully integrated in the entire order fulfilment process of the manufacturer towards dealers and resellers. As a result, supplies of product modules to the assembly line of the manufacturer are 'pulled' and delivered just-in-time (JIT). This has led to suppliers to be highly responsible for the quality and the interface coordination of the product modules supplied and their exact fitting into the car together with the modules delivered by other suppliers. This emphasises the importance of intensified information and knowledge sharing between all suppliers organised by the manufacturer in an open network structure. All knowledge in the network is regarded and maintained as the property of the entire network of the manufacturer and the suppliers. This is particularly important during the development of new products. Suppliers are involved early and intensively in the development, particularly first-tier suppliers.

Aerospace industry, notably Airbus

Aircraft manufacturers like Airbus work with external production locations at suppliers' plants for major subassemblies. For the exact fitting of the subassemblies, suppliers are coordinated via platform strategies. This has led to shorter lead times by increased effectiveness and efficiency of the engineering of the integrated product and the subassemblies. In addition, major suppliers are involved in risk sharing partnerships and cost cutting programmes. An important part of the control includes the logistics and transport of major subassemblies from the suppliers to the manufacturer's plant. Another important part of the control applies to quality control. Dominant quality control by the manufacturer starts during the centrally controlled engineering and the prescription of the engineering software to be used by suppliers. Strict rules and checks for quality assurance apply to the suppliers and the production at the manufacturer as well. As a part of a total quality approach, approval and surveillance of suppliers' capabilities take place on a permanent basis.

Computer industry, notably Dell

Leading computer manufacturers such as Dell have applied direct selling and retailing to end customers by integrating all sales and distribution in their own organisation. Manufacturers have engaged in long-term contracts with suppliers to increase the suppliers' performance, improve service and technological capabilities, and reduce the price of supplies. As a consequence, the supply, assembly and sales of computers have become closely linked. This includes the connections between the assembly lines around the globe. The connection with sales and customers around the globe enables manufacturers to accurately and efficiently forecast production, and match demand with supply on a large scale. Quality standards to suppliers are generally high, and quality failures are not accepted. If components malfunction, suppliers are required to redevelop the components instantly. Structural non-compliance by a supplier is countered with action plans of the manufacturer to improve quality together with the supplier.

Electronics industry, notably ASML

For manufacturers of large electronic appliances, such as ASML, fast and effective channel management has become a key business strategy, not only to increase the speed of production, but more importantly, to reduce the time to market of products with an ever shorter life cycle, such as semiconductor machines. Speed and good forecasting have become essential aims and competitive goals, yet lead

Table 43: Development levels of the applications of supply chain integration found in the building case studies

Factors of analysis	Level 1	Level 2	Level 3	Level 4
	Independent stages	Loosely coupled stages	Closely connected stages	Integrated supply chain
Economic governance				
Integration of business activities	Client 1 Builder 3			Client 2 Project developer Designer 1 Designer 2 Builder 1 Builder 3 Suppliers
Partner sourcing and collaboration strategies		Developer Designer 1 Designer 2 Builder 3	Client 1 Builder 1 Builder 2 Suppliers	Client 2
Production management				
Integration of operations and processes		Client 1 Client 2 Developer Designer 1 Designer 2 Builder 3 Suppliers	Builder 2	Builder 1
Planning and logistics control	Client 2	Developer Designer 1 Designer 2 Builder 2 Suppliers	Client1 Builder 3	Builder 1

Quality management	Client 1 Client 2 Developer Designer 1 Suppliers	Designer 2 Builder 1 Builder 2 Builder 3		
Inter-firm governance				
Information exchange	Client 1 Client 2 Developer Designer 1 Suppliers	Builder 3	Designer 2 Builder 1 Builder 2	
Product development and design		Client 1 Client 2 Designer 1 Designer 2	Builder 2 Builder 3 Suppliers	Developer Builder 1
Market approach and marketing	Client 2 Designer 2	Client 1 Designer 1 Builder 2 Builder 3 Suppliers		Developer Builder 1
Social governance				
Cultural alignment	Client 2 Designer 1 Designer 2 Builder 3	Client 1 Developer Suppliers	Builder 1 Builder 2	
Human resource management	Developer Designer 1 Designer 2	Client 1 Builder 1 Builder 3 Suppliers	Client 2 Builder 2	

times of such complex systems are often long. Efforts are therefore aimed at building trust and transparency with suppliers, although the market and therefore orders are highly volatile. In the development and production, key suppliers are heavily involved in the achievement of those aims. An additional and essential part of supplier management is total quality management and continuous improvement with suppliers to improve technologically complex parts.

Clothing and textile industry, notably Zara

Advanced manufacturers such as Zara have returned to vertical integration as a business model, internalising all business activities in house. Sourcing strategies have mainly been aimed at cost-efficient purchasing of fabric. Delivery processes to the shops are highly integrated to improve responsiveness, increase speed and reduce inventory levels. Conversely, return processes from the shops are highly integrated in order to capture information efficiently about customers' preferences and new trends. All efforts are aimed at speed and strict control of inventory, preventing stock-outs or oversupply. Information exchange is real-time to support logistics and materials supply and to capture market information. Maintaining the quality of products and especially the quality of materials is considered standard routine, and quality goals are set as a straightforward target for suppliers to meet. The quality of design, however, is a very important and highly competitive issue. An important element of market strategies is releasing new products constantly in order to stay ahead of competitors, to create brand loyalty, and to increase the frequency of customer visits to shops.

Food and grocery sector, notably Wal-Mart

In the food and grocery sector, logistics and transport are key issues of supply chain integration, particularly to increase delivery speed and reduce inventories of perishable goods. Establishing cost-efficient purchasing relations with producers and growers, and establishing one's own brands have become essential paths of supply chain approaches for retailers and supermarkets. For instance, Wal-Mart has engaged in strategic partnerships with large manufacturers to reduce cost levels, increase profitability and improve responsiveness. On an operational level, planning and logistics are aimed at quick replenishment, avoiding oversupply, and keeping inventories low. Reverse information is automatically captured by processing sales information and, together with forecasting information, constantly generates new replenishment information. By capturing client information via client programmes, replenishment can be planned even more accurately and individually for each store.

8.2.2 Cross-case analysis between the manufacturing comparators

The comparative analysis is particularly important to search for similarities and differences between the groups of industries explored. The groups are represented by the three supply chain types, each containing two particular industries as categorised previously in Chapter 6: Type 1 *convergent assembly* supply chains, i.e. automotive and aerospace; Type 2 *divergent assembly* supply chains, i.e. computers and electronics; and Type 3 *divergent differentiation* supply chains, i.e. clothing and textile, and food and grocery (Table 44).

Similarities between supply chain integration practices applied in manufacturing

Supply chain integration practices in manufacturing are aimed at reduction of *costs* and *time* to market, improving *customer satisfaction*, maintaining intensive *relations* with supply chain partners and approaching the supply chain as *one channel* that needs to be optimised. The *focal firm* is dominant in the processes down to the supply base as well as up to the end market. In all cases, *market responsiveness* and *operational excellence* are important. And in all cases, manufacturers *push* the market and *create demand* for new products.

Differences between supply chain integration practices applied in manufacturing

Besides the similarities, there are also typical differences between the applications of supply chain integration across the manufacturing cases. In the sectors with a *Type 1* supply chain structure, technological advancement, engineering capacity and product development are centrally organised in tiered supply chain structures mobilising large numbers of suppliers. In sectors with a *Type 2* supply chain structure, fast product assembly is a central issue dealt with by applying modular approaches to end products consisting of interchangeable parts, of which technological function and interfaces need to be optimised in collaboration with key suppliers. In sectors with a *Type 3* supply chain structure, the approach to products and production is characterised as delivery of commodities that need to be brought to a volatile and uncertain market in the fastest and most efficient manner to increase responsiveness and profitability, supported by systems contributing to those aims.

As a further classification, certain issues appear to be of different importance to different supply chain structure types. A strong *commercial focus* is of importance in Type 2 and 3 supply chains, along with being more closely connected to the end customer than in Type 1 supply chains. Therefore, the importance of *speed* per supply chain type is different as well. In Type 1 supply chains, speed is relatively less important than in Types 2 and 3. In Type 2 and 3 supply chains, manufacturers aim to be in *absolute control*. The importance of inventories is different. In Type 1 and 2 supply chains, *inventory control* is of relatively less importance than in Type 3 supply chains. *Real-time information* through the supply chain is more important in Type 3 supply chains. Specialist roles, particularly of suppliers involved in *product development*, are more of an issue in Type 1 and 2 than in Type 3 supply chains. As a result, the same is true for the issues of *quality management*, *design rationality*, *strategic cooperation* and *technological supremacy*. In addition, because of the convergence of supply and assembly and generally larger products, the issue of *teamwork* is particularly important in Type 1 supply chains.

Table 44: Similarities and differences between the findings from the manufacturing comparator studies based on 2nd order themes

Themes	Type 1 supply chain: convergent assembly	Type 2 supply chain: divergent assembly	Type 3 supply chain: divergent differentiation
	Automotive Aerospace	Computers Electronics	Clothing Grocery
Focal firm dominance	Focal firm exercising power mainly towards supply base	Focal firm exercising power across entire supply chain	Focal firm exercising power mainly towards market
Market responsiveness	N/A	Direct sales and fast delivery are key market factors.	Fashionable and perishable goods require fast delivery.
Operational excellence	Operational excellence is a key contributor to cost reduction and profitability.	Operational excellence is a key contributor to cost reduction and profitability.	Operational excellence is a key contributor to cost reduction and profitability.
Market push	Technological innovations push market demand.	Technological and commercial innovations push market demand.	Commercial innovations push market demand.
Commercial focus	Technological focus dominates commercial focus.	In particular, computers/retail require supply chains to be highly commercial.	Retail requires supply chains to be highly commercial.
Speed	Speed of less importance than in Type 2 and 3 chains.	Technological obsolescence requires fast operations.	Shelf lives and fashion trends require fast operations.
Absolute control	Dispersed control to suppliers and locations.	Speed and advanced technology require full control.	Speed and intense competitiveness require full control.
Real-time information	Information management based on forecasting	Information management based on forecasting	Timeliness of distribution requires direct information.
Inventory control	Suppliers reduce inventory to improve control and control costs	Suppliers reduce inventory to improve control and control costs	Restricted shelf lives and fashion trends require low inventories.
Specialist roles	Advanced technology of products and components require capable suppliers.	Advanced technology of products and components require capable suppliers.	Less reliance on advanced technological products, and thus also specialist roles.
Total quality	Composite nature of products requires total quality control.	Composite nature of products requires total quality control.	Quality regarded as a simple requirement.
Design rationality	Composite nature of products requires rational design.	Composite nature of products requires rational design	Commerciality of design more important than rationality.
Strategic cooperation	High-tech nature of products and innovations thrives on collaboration with suppliers.	High-tech nature of products and innovations thrives on collaboration with suppliers.	Main focus on either market or internalisation.
Technological supremacy	Technologically supreme products and innovations are key to sustained competitiveness.	Technologically supreme products and innovations are key to sustained competitiveness.	Less reliance on advanced technology.
Teamwork	The manufacture of large, complex products in particular requires supply chain teamwork.	Particularly in the design, teams of supply chain participants apply.	Reduced amount of teamwork among the supply chain participants.

Table 45: Development levels of the applications of supply chain integration found in the manufacturing comparator studies

Factors of analysis	Level 1	Level 2	Level 3	Level 4
	Independent stages	Loosely coupled stages	Closely connected stages	Integrated supply chain
Economic governance				
Integration of business activities			Airbus ASML	Toyota Dell Zara
Partner sourcing and collaboration strategies			ASML Zara Wal-Mart	Toyota Airbus Dell
Production management				
Integration of operations and processes				Toyota Airbus Dell Zara
Planning and logistics control			Airbus ASML	Toyota Dell Zara Wal-Mart
Quality management		Zara	Toyota	Airbus Dell ASML
Inter-firm governance				
Information exchange		ASML		Toyota Airbus Dell Zara Wal-Mart
Product development and design			Toyota Dell	Airbus ASML Zara
Market approach and marketing			ASML	Toyota Dell Zara
Social governance				
Cultural alignment			Dell ASML	Toyota
Human resource management				Toyota

Per factor of analysis the *development levels* have been grouped based on the observation of the various supply chain integration practices in the different comparator studies (Table 45). The differences in their presence and appearance also show variance in the levels between the industries. Although the levels are indicative and therefore not absolutely comparable, levels of supply chain integration are relatively high. This particularly concerns the relations between retail, manufacturing and supply. Often it includes product development and design. In a few cases the supply chain integration strategies have led to vertical integration i.e. internalisation of activities, e.g. Zara.

8.3 Comparative analysis between building and manufacturing

Finally, based on the results of the respective analyses of the building case studies and the manufacturing comparator studies above, the findings from those analyses are compared, in particular to identify characteristic *similarities and differences* between both (Table 46).

Comparing building to Type 1 manufacturing *convergent assembly*, notably automotive and aerospace

The building industry is probably best comparable to Type 1 manufacturing, since both share the characteristic of *convergent production*, often combined with a *make-to-order* approach to product delivery. *Platform strategies* apply to both sectors. Both also make use of *distributed production* locations of parts, typically *prefabricated components* produced at suppliers' plants. However, differences can be seen in the approach to *product development* of integrated parts by suppliers, and the structural approach to *first-tier suppliers* managing lower-tier suppliers, which is customary in the automotive and aerospace industries, but not in building. The *centralised engineering* of products by manufacturers in the automotive and aerospace industries is different from the decentralised outsourcing of the engineering generally applied by contractors in the building industry.

Comparing building to Type 2 manufacturing *divergent assembly*, notably computers and electronics

Type 2 manufacturing seems less comparable to building. However, in Type 2 manufacturing as well as in the building industry, *modular approaches* to design and production are applied, although in building it has been considerably less widespread and not based on industry standards as in the computer and electronics industries. Typically different is the *market approach* which has been characterised in both sectors of manufacturing previously mentioned, that relies on *direct sales* to end customers and *fast channel management* aimed at reduced time to market. The *fast obsolescence* of products and technologies in both sectors of manufacturing is a key driver for such approaches, which is a fundamental difference from the building industry.

Comparing building to Type 3 manufacturing *divergent differentiation*, notably clothing and grocery

Type 3 manufacturing is probably the least comparable to the building industry. However, two aspects may be comparable. First, the approach of *design and customisation* of products is similarly based on trend watching, translating this into a design by a designer and sometimes allowing individual client wishes to a certain extent. Similarities can also be seen in the attention paid to *logistics and transport* of products and supplies to or from geographically dispersed sites. However, both in clothing and grocery this refers to the often *globally* organised flow of final products to the end customer. In building, however, this refers to *locally* bound materials supplies shipped to a construction site. Another typical difference is the *direct sales* and short *connections to the end customer* in both sectors in contrast to building.

8.4 Reflection on key issues and contributions of this chapter to the thesis

When comparing the applications of supply chain integration in manufacturing to those in the building industry, fundamental differences appear. The applications in manufacturing appear to be rather comprehensive and advanced, and apparently follow mainstream theory in the area of supply chain integration. In contrast, the applications in the building cases seem rather industry-specific, including pragmatic translations of manufacturing practices adapted to the production situation in the building industry. The scope and level of the applications in the building cases also vary depending on which firm type represents the focal firm. This situation is relatively ambiguous in contrast to manufacturing where it is clear that the lead manufacturer is generally the focal firm. Hierarchical relations and roles between firms in the manufacturing supply chain are generally clearer. In addition, the factors of analysis are of equally broad importance in manufacturing supply chains, in contrast to building supply chains (Table 47).

In general, the explanation for these differences can first be found in the higher level of maturity and dissemination of the phenomenon of supply chain integration across manufacturing in contrast to the building industry. The understanding and explanation of the aspects of supply chain integration in manufacturing are quite unambiguous. In particular, the aspect and level of centralisation exercised by the focal firm is generally undisputed in manufacturing supply chains in contrast to building supply chains. Additionally, the level of repetitiveness is typically high in manufacturing supply chains, and in manufacturing industries in general, extending to virtually all aspects of products, processes, firm structures and interfirm collaborations. The building cases in this research generally showed higher levels of centralisation indeed. The issue of repetitiveness appeared to be less or only implicitly present. However, to be successfully applied, it is likely to be an important factor for supply chain integration in building, too. Acknowledging the importance of repetitiveness, it has been added as an eleventh factor of analysis.

Table 46: Cross-case analysis between the main findings from the building case studies and the manufacturing comparator studies

Factors of analysis	Building	Manufacturing	Similarities	Differences
Economic governance				
Integration of business activities	• In building, firms have an apparent preference for internalising business activities.	• Manufacturing has generally moved away from totally internalising business activities.	• Focal firms have a general urge to control selected business activities across the supply chain.	• In manufacturing, it mainly applies to manufacturers aimed at market power of the supply chain.. • In building, it applies to various types of firms, including the client aimed at influencing the power balance in the supply chain.
Development level	1-4	3-4		
Partner sourcing and collaboration strategies	• Long-term but less comprehensive collaboration, relatively disconnected, fewer activities shared than are possible.	• Structured and strategic approach to collaboration with particular suppliers, as external business units. • Dominant role is played by the focal firm.	• Basically the same long-term approach to inter-firm relations exists.	• In manufacturing, this is mainly aimed from manufacturer to suppliers, typically intensively. • In building, in exists across the supply chain among various firms, often loosely.
Development level	2-4	3-4		
Production management				
Integration of operations and processes	• Rather loosely coupled operations and processes, per project. • Production control dispersed among firms, in particular the builder.	• Strict cross-corporate alignment of operations and processes. • Leading production control by focal firm towards supply chain.	• Focal firms aim at increased alignment of operations and processes.	• Alignment and production control much stricter in manufacturing than building. • In manufacturing, overall production control is centralised at the focal firm.
Development level	2-4	4		
Planning and logistics control	• Joint project planning between supply chain firms. • Integrated planning of site logistics and operations.	• Sales and production planning, inventory control are integrated as part of strictly controlled production system • Planning for speed.	• Focal firms aim for joint planning with supply chain firms.	• In manufacturing, planning and logistics integrated from sales to supply, using advanced systems. • In building, restricted to site work, per project.
Development level	1-4	3-4		
Quality management	• Quality management dispersed along supply chain • Quality often seen as customer satisfaction, subjective, relative.	• Total quality management across supply chain processes. • Non-acceptance of quality problems, absolute compliance.	• Firms in the supply chain are reliant on each other's quality being delivered.	• In manufacturing, quality is defined and managed comprehensively and uniformly across supply chain, requiring absolute compliance from all firms. • In building, quality is relative, and the quality management is dispersed across the supply chain.
Development level	1-2	2-4		

Inter-firm governance				
Information exchange	• Generally loosely coupled information exchange, in some cases supported by BIM or Internet portals.	• Advanced information sharing, real time, entirely from supply base to retailers and back.	• Firms in the supply chain share urge to exchange information.	• In manufacturing, all information is shared systematically across the supply chain to increase speed. • In building, information sharing is generally rather poorly organised.
Development level	1-3	2-4		
Product development and design	• Design generally isolated by separate designer. • In some cases with developer or builder, in house at builder's company, or designer also builds.	• Product designs are modularised, rationalised. • Standardised, interchangeable parts, with suppliers involved.	• Focal firms try to integrate product design, development in the process.	• In manufacturing, design is integrated part of production, based in standardisation, modularisation. • In building, firms try to control design by ownership, e.g. BIM, but are not necessarily rationalising design.
Development level	2-4	2-4		
Market approach and marketing	• Sales channel fragmented with many intermediate parties. • Marketing as a separate activity is virtually absent. • Clients are often the focal firm, and directly involved in the supply chain and in the project operationally.	• Integrated sales channel controlled by the focal firm. • Short time to market, and high market responsiveness. • Market approaches specifically aimed at customers, but not involving them in the supply chain.	• Firms in the supply chain try to shift and move towards end market, client.	• In manufacturing, focal firm leads marketing, by integration of sales channel, direct selling, increasing market responsiveness with suppliers. • In building, sales channel is mostly fragmented, and the producer disconnected from the end market.
Development level	1-4	3-4		
Social governance				
Cultural alignment	• Social and informal links between firms and individuals. • Different corporate cultures, values persist. • Low levels of collective organisational culture.	• Supplier development programmes in place. • Focal firm aims for high level of collectivism, shared values. • Common understanding of organisational culture.	• Focal firms aim for increased mutual understanding between firms within the supply chain.	• In manufacturing, formalised effort is invested in cultural alignment, collectivism, shared values. • In building, different corporate cultures and informal links persist.
Development level	1-3	3-4		
Human resource management	• Joint working in boards, engineering and project management. • Informal knowledge sharing.	• Formally organised personnel exchange. • Formalised knowledge sharing.	• Firms in the supply chain use joint teams with other firms.	• In manufacturing, formal measures are taken for personnel and knowledge exchange. • In building, joint working and exchange mainly via daily activities, informally.
Development level	1-3	4		

Table 47: Reflecting on the importance of the ten factors for supply chain integration in the building industry versus manufacturing

Factors of analysis	Building	Manufacturing
Economic governance		
Integration of business activities	In most cases the focal firm at least partly achieves supply chain integration via internalisation of selected activities.	Manufacturing firms have generally outsourced activities to strategic partners, except for activities towards the market, which have been internalised.
Partner sourcing and collaboration strategies	Strategic collaboration appears to be only selectively established with varying types of firms and generally does not cover the entire supply chain.	Strategic collaboration has been established with virtually all firms in the supply chain, particularly towards the supply base.
Production management		
Integration of operations and processes	Operational integration of processes among firms is uncommon.	Operational integration of processes among firms is common.
Planning and logistics control	Planning is often a joint activity, but logistics has not often been integrated or strictly organised.	Planning and logistics are typically approached jointly and strictly organised with suppliers and third party service providers.
Quality management	Quality management is seldom an explicit topic of supply chain integration.	Total quality management is an integrated part of supply chain integration.
Inter-firm governance		
Information exchange	Information exchange is generally limited to the company's own organisation and the project, and is often informally organised.	Information is extensively shared within the supply chain, and formally organised.
Product development and design	In the case of internalisation, the particular firm generally also develops repetitive product formats, instead of project design. However, in all cases the product needs adaptation to project specifications.	Product development is a strategic issue and organised closely to the focal firm, generally involving suppliers in the development. Manufacturability and rationality of the design are key issues.
Market approach and marketing	In a few cases, firms have managed to deliver directly to the end market.	Focal firms either own or control the sales channel as a key factor of commercial success.
Social governance		
Cultural alignment	In a few cases, suppliers in particular have been aligned, however informally.	In a manufacturing setting, confirmation of the entire supply chain to the culture of the focal firm is a formalised and univocal procedure.
Human resource management	Generally, joint working is restricted to daily activities in project environments; in a few cases collaborative working has been organised structurally.	Joint working and personnel exchange is standing procedure and organised actively and formally, e.g. via personnel exchange programmes.

In the following chapter, for each of the eleven factors of analysis, the empirical findings from the building case studies will be confronted with theory, and using the findings from the manufacturing comparator studies as bridges between both. Following Eisenhardt (1989), the aim here is to identify and clarify relations between theory and practice by iterating between both. As a result, hypotheses are shaped and contributions to existing theory are made, adding to the concept of supply chain integration for the building industry at which this thesis is aimed.

127
128
130
131
BONIPAK
Fresh Premium Broccoli
CROWN PACK
PRODUCE OF USA
www.bonipak.com
BONIPAK
BROCCOLI
FRESH BROCCOLI
FRESH BROCCOLI
FRESH BROCCOLI
BONIPAK

9 Conclusions

This thesis has endeavoured to contribute to theory for supply chain integration in the building industry, based on existing theoretical concepts and applications of supply chain integration in manufacturing and in building. This chapter finalises the theory-building process of this thesis by shaping hypotheses, thus representing the closure of the research process following Eisenhardt (1989). The main result constitutes the thesis' contribution to theory, i.e. the hypotheses reflecting the concept of supply chain integration for the building industry. In addition, the chapter answers the research questions that led to the theory building. Next, the theory building approach itself is discussed, notably the use of Eisenhardt's case study method for theory building. In addition, the implications of the thesis' outcomes and their potential consequences for theory and practice are mapped out. To conclude, optional directions for further research will be given, including suggestions for the advancement of theory, method and practice beyond this thesis.

9.1 Theory building: confronting theory with practice

For the finalisation of the theory building process, theory is confronted with practice, i.e. *literature* is examined in the light of the *empirical findings* from the building case studies, and vice versa, following Eisenhardt's methodology for theory

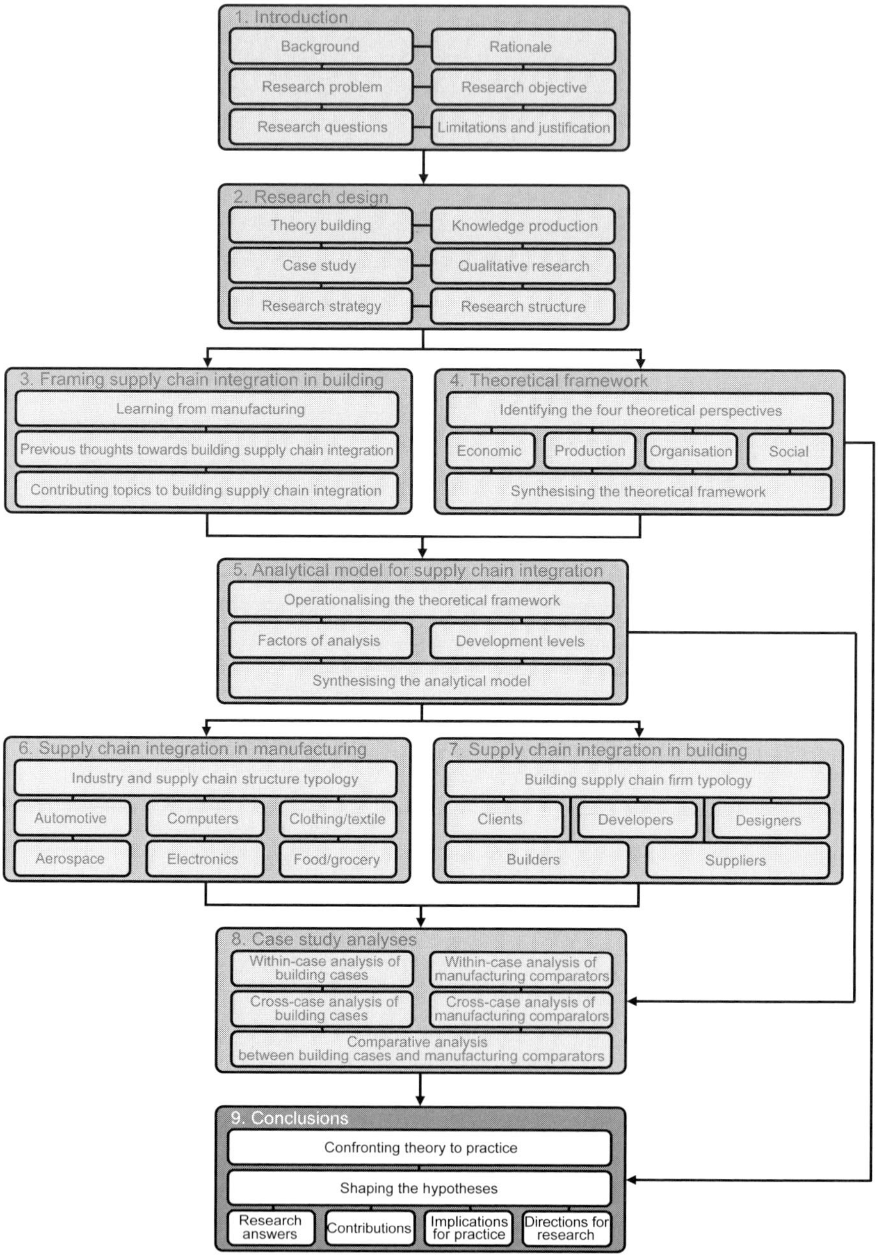
1. Introduction
Background
Rationale
Research problem
Research objective
Research questions
Limitations and justification
2. Research design
Theory building
Knowledge production
Case study
Qualitative research
Research strategy
Research structure
3. Framing supply chain integration in building
Learning from manufacturing
Previous thoughts towards building supply chain integration
Contributing topics to building supply chain integration
4. Theoretical framework
Identifying the four theoretical perspectives
Economic
Production
Organisation
Social
Synthesising the theoretical framework
5. Analytical model for supply chain integration
Operationalising the theoretical framework
Factors of analysis
Development levels
Synthesising the analytical model
6. Supply chain integration in manufacturing
Industry and supply chain structure typology
Automotive
Computers
Clothing/textile
Aerospace
Electronics
Food/grocery
7. Supply chain integration in building
Building supply chain firm typology
Clients
Developers
Designers
Builders
Suppliers
8. Case study analyses
Within-case analysis of building cases
Within-case analysis of manufacturing comparators
Cross-case analysis of building cases
Cross-case analysis of manufacturing comparators
Comparative analysis between building cases and manufacturing comparators
9. Conclusions
Confronting theory to practice
Shaping the hypotheses
Research answers
Contributions
Implications for practice
Directions for research

building (Eisenhardt 1989). This final part of the theory building process implies the shaping of *hypotheses*, which result by *relating* the theory presented in Chapters 3, 4 and 5 to the findings from the building case studies presented in Chapter 7. The manufacturing comparator studies presented in Chapter 7 function as an *intermediate* step in the confrontation between the theory and the building case studies, adding to the explanation of supportive *similarities* or conflicting *differences* between the theory and building practice, and thus giving additional direction to the theory building process.

The shaping of the hypotheses involves identifying and verifying *relationships* between the theory and the empirical findings following from the thesis (Eisenhardt 1989). The issue here is that the *emergent theory* must closely fit the findings uncovered in practice. In order to ensure this, first the main issues from the literature presented in the *theoretical part* of this thesis will be reviewed, and the relative importance of their separate contributions to the concept of supply chain integration. This body of theory will next be related to the main findings derived from the *empirical part* of this thesis. The process of finding those relationships implies an *iterative comparison* between theory and empirical findings, leading to a set of hypotheses further in this chapter. In this process we use *replication logic* of the findings from the case studies confirming or disconfirming the theory, and thus shaping the resulting hypotheses (Yin 1989). Empirical findings *confirming* the theory enhance the confidence in the validity of the hypotheses. Empirical findings *disconfirming* the theory provide an opportunity to extend and refine the hypotheses (Eisenhardt 1989).

The method of shaping the hypotheses follows a path of multiple *tabulation* and *reduction* (Eisenhardt 1989). First, the implicit relationships between factors and goals of supply chain integration found in the preceding parts of the thesis will be identified. Next, both the factors and the goals will be reduced. The *reduced factors* and the *reduced goals* will finally be related and rephrased to the resulting hypotheses.

9.1.1 Identifying the relationships supporting the hypotheses

The *relationships* between the *factors* and *goals* of supply chain integration follow from the thesis as it unfolded in the previous chapters. These relationships follow from the ten factors of analysis of the analytical model used to analyse the case studies, plus the eleventh factor that emerged from the case study analyses, namely repetitiveness. For each of the *eleven factors*, we have summarised and confronted the findings from theory and practice in the series of condensed tables below. The theory is displayed on the left side of the tables, building practice on the right side, and manufacturing practice as a bridge in the middle. This has led to the identification of *key relationships* between theory and practice, process and product implications, and means and goals of supply chain integration.

Factor 1: Repetitiveness

The concept of repetitiveness and finding or developing concepts and methods that increase the level of repetitiveness is a key factor of supply chain integration. In the case of building, repetitiveness between projects, rather than delivering one-off projects, has appeared to be an issue for supply chain integration (Table 48).

Table 48: Factor 1: Repetitiveness

Theory	Manufacturing practice	Building practice
• Repetitiveness is referred to as discovering or developing concepts and methods that increase the level of repetitiveness of products and processes. • Typically the focal firm is leading and dominant in the initiation and promotion of repetitiveness thro ugh the supply chain regarding products and processes as well as organisational formats. • In construction management literature, increased repetitiveness between building projects has been achieved by introducing and increasing the repetition of integrated production and business concepts. • Focal firms need to apply long-term approaches to the marketplace, design, engineering, parts manufacture, logistics and site assembly, rather than delivering one-off projects based on mere project specifications.	• Increased repetitiveness in processes and products which has been linked to large investments in inter-firm relations, resources and innovations, has constituted a great deal of the success of firms and the supply chain integration strategies applied. • From the position of the focal firm, strategies upstream in the supply chain towards the supply base have mainly been aimed at increasing the repetitiveness of processes and relations with key suppliers. • Downstream, the supply chain repetitiveness is particularly sought in product formats and market formulas.	• Comparable approaches exist, including higher levels of investments, and also higher levels of repetitiveness, particularly as opposed to competitors and the wider industry tradition of a one-off project view on products and production. • However, unlike manufacturing and theory, repetitiveness in building is not fully applied coherently to all aspects mentioned, often only to the next stage in the supply chain, and is mainly restricted to fixed relations. • Focal firms tend to exercise their influence to promote repetitiveness in all aspects only to a limited extent. The approaches are aimed flexibly at multiple projects rather than structurally towards repetitiveness. • Repetitive approaches to processes have seldom been applied and to markets and products even less, mainly because of building's restrictions on applying repetitive approaches in those directions.
Relationship		**Leads to**
A coherent approach to increased repetitiveness of products, processes and inter-firm relations upstream and downstream in the supply chain		Improved performance in terms of products and processes, organisations and individuals, productivity of inputs and quality of outputs.
Process implications		**Product implications**
Efficiency towards the supply base is improved due to repetitive relations and processes.		The market effectiveness is improved due to repetitive products.
Means		**Goals**
• Increased repetitiveness of products and parts. • Increased repetitiveness of organisational and process formats.		• Effective and responsive market activities. • Efficient operations and processes.

Factor 2: Integration of business activities

The level of integration of business activities in the supply chain is essential. In the case of supply chain integration, the focal firm has integrated internal business activities, and possibly external business activities as well. This depends on the make-or-buy decision of the focal firm per business activity (Table 49).

Table 49: Factor 2: Integration of business activities

Theory	Manufacturing practice	Building practice
• The level of integration of business activities within the supply chain is an essential issue for the focal firm. The focal firm needs to consider and balance integrating internal business activities versus controlling external business activities. • For the integration of external activities, the focal firm needs to balance whether to take financial control or ownership over those activities, or engage into decentralised control over business activities outside their own firm's boundaries. • Financial ownership leads to vertical integration, which may at a certain level cease to be beneficial because of negative side effects. Various strategies towards vertical integration are possible, such as quasi-integration, implying a lower level of integration. • The idea of the 'quasi-firm' has raised the issue of core competences of firms together making up an 'extended enterprise', or a 'virtual organisation'.	• Vertical integration has gone generally out of fashion, replaced by virtual integration, i.e. decentralised control of suppliers. • Decentralised control is a substitute for internalisation, consisting of strict production and organisational compliance from the side of the suppliers to the focal firm via concepts such as platform strategies and modular networks. • However, core competences as well as sales and market activities have generally been internalised firmly for strategic reasons such as close contact with markets and customers. • In volatile markets, integration efforts have been aimed at speed and fast channel management. In such conditions, manufacturers have rediscovered vertical integration, to cope with volatility and uncertainty of client behaviour.	• Internalisation has been commonly found, but efforts have mainly been aimed at controlling neighbouring stages in the supply chain, and often not towards the end market. • This includes clients shifting activities to contractors or entering into legal partnership with contractors. • Another path is integrated delivery of built assets and all services by one firm. A last direction observed was internalising or teaming up various business functions in order to deliver integrated products or services.
Relationship		**Leads to**
Integration of selected business activities by the focal firm within the organisation through internalisation and increased decentralised control.		Increased power and control over activities within the focal firm together with the firms responsible for external activities.
Process implications		**Product implications**
A larger part of the processes in the supply chain comes under direct or indirect control of the focal firm.		Increased branding of the product by the focal firm.
Means		**Goals**
• Integration of business activities by focal firm internally. • Decentralised control of external business activities.		• Improved market position of the focal firm. • Improved competitiveness of the focal firm.

Factor 3: Partner sourcing and collaboration strategies

Partner sourcing is essential to find the right partner firms. Because strategic collaboration with those firms needs to be put in place, procurement and purchasing practices need to be shifted to a strategic level. Strategic collaboration in general takes a long-term perspective to contractual relationships (Table 50).

Table 50: Factor 3: Partner sourcing and collaboration strategies

Theory	Manufacturing practice	Building practice
• Besides integration of business activities, a focal firm will have to establish strategic arrangements and contracts with partner firms in the supply chain, typically suppliers contracted based on a long-term basis. • Strategic collaboration with the suppliers needs to be put in place. Collaborative contractual relationships imply non-adversarial bargaining and relational contracting. • In a building context, the movement has been towards establishment of partnering and framework arrangements within the supply chain. • Application of strategic procurement approaches has led to multi-project approaches towards integrated clusters of firms taking care of those projects for longer periods of time, e.g. integrated life-cycle contracts. • Contractors are viewed as partners adding value to the client's business and helping to drive through cost reductions on a structured basis instead of merely doing projects for the client.	• Strategic sourcing typically has taken place between the focal firm, i.e. manufacturer, in close relation with suppliers organised in a tiered structure of external business units. • Manufacturers have reduced their supply base, entering into long-term contracts with key suppliers. In particular, first-tier suppliers have been involved and responsible for well-defined packages of activities and strict delivery and quality conditions of integrated parts to the manufacturer, including such issues as risk sharing, cost cutting, increase of profitability and responsiveness. • In addition, the first-tier suppliers have been held responsible for the quality and product development of the integrated parts they supply, and involving and managing lower-tier suppliers along with it. • Supply management is reflected in a modular approach to supply chains and production networks.	• Various firm types have entered into fixed, long-term relations with various parties supplying different kinds of services, products or parts. • These have primarily been bilateral arrangements such as between clients and contractors, or contractors with suppliers. Most of the arrangements have been narrowly aimed at cost and time reduction, and a few more widely aimed at joint design and production. • Clients have typically moved towards the use of partnering and framework agreements to rationalise their project approach and their supplier base along with it. • Clients have created multi-project environments implying multi-project procurement and repetitive tendering arrangements with contractors.
Relationship		**Leads to**
Establishment of strategic collaboration and long-term arrangements by the focal firm with other firms in the supply chain.		Improved organisation, commitment and involvement of those firms in the supply chain on an operational level as well as on a strategic level.
Process implications		**Product implications**
Sourcing and purchasing strategies are shifted to partnering arrangements.		Products explicitly become joint products produced and delivered collaboratively by all firms in the supply chain.
Means		**Goals**
• All firms in the chain are involved in partnering arrangements. • Long-term, strategic and intensive collaboration.		• Improve collective market position. • Improve competitiveness of chain as a whole.

Factor 4: Integration of operations and processes

Process integration is a key aspect of supply chain integration aimed at 'flow of production'. The flow of production particularly refers to 'flows' of material and operations, which are 'pulled' through the process (Table 51).

Table 51: Factor 4: Integration of operations and processes

Theory	Manufacturing practice	Building practice
• Relations and interfaces between processes and operations in the supply chain need to be strictly organised and integrated in order to strive for flow of production. • The flow of production particularly refers to flows of material and operations. Typically flow of production is referred to in lean production, and as in lean construction implies intensified coordination of the supply base, particularly to resolve waste and buffers and increase efficiency and productivity. • In contrast to lean production, agile production is focused on increased effectiveness and flexibility of production and the pursuit of a higher connection and responsiveness of the production output towards the market and customers. • The logic of lean and agile production as applied in manufacturing is based on the idea that supply chain management is more effective with stable supply chains and with 'mass-customised' products to achieve the goals of both increased efficiency (lean) and greater effectiveness (agile).	• The integration of processes and operations to achieve flow production has been observed to be standard practice. Suppliers have been fully integrated in the order fulfilment process of the focal firm all through supply, assembly and sales. • Integration of suppliers is based on strict alignment of process interfaces and central production control by the focal firm. Integrated order and inventory management is part of the production system. • Suppliers are treated as external business units. Global activities are also connected. Processes are highly integrated for speed, inventory control, and information control. • The involvement of suppliers is generally aimed at achieving high levels of effectiveness and efficiency of operations and processes.	• Similar levels were approached in the case of complete, in-house integration of all operations in a factory. Lower levels of integration of processes and operations have also been found. • Examples however tend to result in business integration rather than operational integration. Joint engineering has been observed, however, as operational integration. • In order to stabilise the supply chain, and hence the making of products together with suppliers, co-maker ship as originated in manufacturing alongside lean and agile production, has found its way to building. • On-site operational integration has been found in applications of lean planning and joint site construction.
Relationship		**Leads to**
Seamless integration of operations and processes through the supply chain.		Improved organisation and control of relations and interfaces between activities of firms involved.
Process implications		**Product implications**
The separate corporate processes in the supply chain need to be united into a single process.		The product produced through the supply chain is a supply chain product, rather than an assembly of parts.
Means		**Goals**
• All firms joint in a uniform process organisation. • Seamless integration of all operations.		• Improved flow of production. • Increased productivity.

Factor 5: Planning and logistics control

Planning and logistics are basic elements of supply chain management. Particularly just-in-time logistics are a part of integrated supply chains. This kind of logistics relies heavily on advanced forms of inventory management and the use of advanced ICT systems (Table 52).

Table 52: Factor 5: Planning and logistics control

Theory	Manufacturing practice	Building practice
• For the integration of processes and operations, integrated planning and logistics are a central issue for supply chain integration, particularly from supplier to manufacturer, and from manufacturer into sales channels. • Just-in-time logistics are a part of integrated supply chains. This kind of logistics relies heavily on advanced forms of inventory management. • The use of advanced ICT systems for joint planning and the use of logistics service providers is part of planning and logistics solution.	• Planning and logistics are greatly advanced, applying sophisticated ICT and 'pull systems' through the pipeline, both from suppliers to the manufacturer, and from the manufacturer to sales. • The assembly lines of the suppliers are 'pulled' (kanban) and supplies delivered just-in-time (JIT). Logistics is part of the design and engineering activities • Shorter lead-times are achieved by joint engineering with suppliers aimed at Design for Manufacturability (DfM). • Sales and resource planning are integrated for accurate forecasting and matching supply to sales. Planning is aimed at reduced lead-times and time to market. Strict control of inventory is an essential element for such aims as lowering inventory, preventing stock-outs or oversupply. • In retail sectors, quick replenishment is supported by distribution centres and logistics service providers.	• Planning and logistics have been found on a lower level, and initiatives were mainly aimed at organisation of the work of contractors with subcontractors and suppliers based on a lean approach. In the cases of the clients, planning generally had to do with high-level planning of projects. • Initiatives have been taken in the field of integrated logistics, although generally not to the level of manufacturing. • The introduction of just-in-time logistics, as part of the concept of lean construction, has resulted in shorter delivery times and lower inventories on building sites. • In none of the building cases were either advanced ICT or pull systems put in place as part of sophisticated planning and logistics systems. In a few cases, BIM applications were used for planning as a side effect.
Relationship		**Leads to**
Integrated planning and logistics across the entire supply chain,		Improved flow of materials, products and services towards the end market, and the reverse flow of demands, orders and information towards the supply base.
Process implications		**Product implications**
All processes across supply chain interconnected and supported by one planning and logistics system.		Product quality and configuration need to be at a high level to support smooth planning and logistics.
Means		**Goals**
• Increased predictability of planning. • Increased responsiveness across the supply chain from the supply base to sales.		• Improved flow of production. • Reduced time to market. • Improved market responsiveness.

Factor 6: Quality management

An integrated approach to quality is an important and basic factor in integrated supply chains. Firms involved in an integrated supply chain such as in manufacturing typically apply total quality management across their organisational boundaries and across their separate production systems (Table 53).

Table 53: Factor 6: Quality management

Theory	Manufacturing practice	Building practice
• Total quality management is a cornerstone and basic requirement of supply chain integration. • Firms involved in an integrated supply chain typically apply total quality management to assure a high level of quality and a minimal level of quality problems across all organisational boundaries and the collective production systems along the entire supply chain. • The lean and six sigma philosophies in particular have advocated continual review and improvement of quality. Relationship stability and integration have been shown to be essential factors to ensure consistently high quality and continuous improvement of quality. • Significantly different are the approach and level of quality management in building.	• Integrated approaches to quality along the supply chain have been observed to be crucial. Joint achievement of high quality by manufacturer with suppliers. • Quality standards have been predefined on a high level for every firm in the supply chain to meet. Nonconformity to the quality levels is not accepted and calls for redevelopment of parts and products, and a review of processes and instant prevention measures. • Suppliers are responsible for the quality of their own parts. Strict rules and checks for quality assurance have been put in place by the focal firm. In addition, the focal firm supervises suppliers' quality management systems. • Quality control starts in the design and engineering phases. Total quality management and continuous improvement are aimed at improving parts and products with suppliers.	• In virtually none of the cases was quality a comparable issue of prominent concern with supply chain partners, nor were joint quality systems in place as part of a supply chain approach. • The attitude tends to be oriented towards compliance with contractual specifications and not gaining additional financial benefits or competitive strength from quality improvement. • Quality assurance is regarded as a collective task in the design phase but as a responsibility of the contractor in the construction phase. • In the case of the integrated house building, quality control took place in the factory. However, quality control was generally aimed at the site installation, and as a final check of conformity to client's wishes. • In a few cases, quality was collaboratively assured via Building Information Modelling (BIM).
Relationship		**Leads to**
Application of total quality management across all corporate boundaries in the supply chain.		Fosters a collective quality approach to improve the absolute quality of the products and services delivered through the supply chain to the end customer.
Process implications		**Product implications**
Total quality system through the chain is firstly aimed at harnessing objective quality.		Basic condition and aim is the objective and technical product quality, followed by the relative quality.
Means		**Goals**
• All firms need to comply with absolute quality rather than relative quality. • Long-term arrangements support continuous improvement.		• First, improved objective quality of the product. • Second, improved relative quality experienced by the customer. • Improved total and continuous quality.

Factor 7: Information exchange

Firms engaged in integrated supply chains must sustain increased levels of information sharing and shared knowledge between firms in and beyond the supply chain. The aim is to be collectively and constantly informed about the status of the supply chain, and to be able to mobilise the information and knowledge needed for existing and new business (Table 54).

Table 54: Factor 7: Information exchange

Theory	Manufacturing practice	Building practice
• Firms in integrated supply chains are committed to information sharing and shared knowledge management between firms in and beyond the supply chain. • Firms may either rely on joint information systems, or they may foster the flow of knowledge and information between firms. • Organisations have been described as 'the coordination of efforts of people working on a collaborative task broken down in a set of specialised activities. Coordination is then achieved through communication'. Next, communication needs to lead to commitment. • With respect to commitments, firms need to manage communication across firms, whether they be closely connected in the integrated supply chain, or loosely coupled in the permanent network of firms within an industry. • Intensified and close communication is a basic premise and an automatic condition for supply chain integration.	• Information exchange and communication have been highly and formally organised by the focal firm and widely spread across the supply chain and sales channels, with suppliers and resellers highly committed and bound to those information flows. This encompasses the boundaries of the supply chain to the wider industry, particularly for global sales, operations and supplies. • Advanced electronic information systems support the information flow. Online engineering and e-commerce applications support the information and knowledge exchange during design and purchasing. • During production, direct information is passed from customers to assembly and suppliers to increase market responsiveness. Real-time information supports logistics, materials supply and transport of the final product to the end customer. • Reverse information systems are put in place to capture information about sales, customer behaviour and trends. • All firms must commit to the information system of the focal firm.	• Information management has shown to be far less structured and integrated. Even in the case of integrated house building, information management was organised informally. • Information management in building is particularly connected to the rather classical approach to the project management function in building. • More advanced approaches to information exchange in building often refer to applications of ICT. • In the cases where Building Information Modelling (BIM) was applied, information exchange was organised in a more structured way via the functionalities of the BIM system.
Relationship		**Leads to**
Commitment of firms to an integrated and transparent information flow through the supply chain.		More effective and efficient information sharing and improved collective working.
Process implications		**Product implications**
The entire supply chain process is supported by a single information system.		The product is translated into a consistent information package.
Means		**Goals**
• All firms connected to a single information system. • Absolute commitment and reliance on information system.		• Information is a single source to align all firms. • Improved joint competitive strength of the firms involved.

Factor 8: Product development and design

Firms in the integrated supply chain are involved in joint product development and design. The firms pool resources along the supply chain, applying techniques such as design for manufacture, early supplier involvement and product modularisation to ensure the connection between design and production capability (Table 55).

Table 55: Factor 8: Product development and design

Theory	Manufacturing practice	Building practice
• Design is part of production, and an integrated function of the focal firm, often the manufacturer, who involves suppliers. • As part of a joint product development and design approach, firms pool resources along the supply chain, e.g. applying techniques such as design for manufacture to ensure the connection between design and production capability. • The technological expertise of suppliers is mobilised during the design by applying the concept of early supplier involvement. Product modularisation is often applied to enable the distributed production of parts or components of the final product. • Product modularisation and distributed production are supported by application of platform strategies, developing a large set of interchangeable parts and components, and thus being able to assemble different product types. • In a building context applications of distributed production imply transferring on-site activities off site, and prefabrication.	• Product development and design have been dominated by the focal firm, based on platform strategies, standardisation and interchangeability of parts, and intensive supplier involvement. • The central idea is to integrate suppliers' processes as early and intensively as possible in the design, engineering and manufacture of parts. • Design for manufacture, i.e. manufacturability is a key issue, and thus a major concern and requirement of product development with suppliers. • Speed of development is important to reduce time to market, particularly for products with short life cycles and volatile markets. New designs are often based on reconfiguring existing products to decrease investments and reduce lead-time. • Manufacturers connect closely with the market to develop new products, involving multi-functional design teams. • Postponement is applied to assure products meet the latest trends and customer demands.	• Design and production have been connected, particularly when designer and contractor are integrated or teamed up, or when either party controls the other stage, i.e. a designer as management contractor, or a contractor developing a product. • Examples include joint project development by clients with contractors, contractors with suppliers, or own integrated product development by developers, architects, contractors or suppliers. • In the case of the suppliers, they have predeveloped and pre-engineered integrated products by pooling suppliers' capacities together and standardising components to the maximum extent. • Reuse of components and increase of prefabrication of larger integrated components have been observed to rationalise product design.
Relationship		**Leads to**
Product development based on the integrated contribution of the commercial, design, engineering, production and supply functions in the supply chain.		Increased product rationalisation, modularisation and standardisation.
Process implications		**Product implications**
The process leads the product development and design, sparking the subsequent activities and actors in the supply chain.		Cost-efficient and commercially successful product formats must result.
Means		**Goals**
• Integrated design and production. • Mobilisation of suppliers' expertise and novel technology.		• Improved manufacturability and quality of products. • Increased commercial success of products.

Factor 9: Market approach and marketing

Firms engaged in an integrated supply chain are subsequently also engaged in developing a joint market approach. Typically, this process is led by the focal firm, involving suppliers but notably resellers and other firms in the sales channel to co-develop and commit to a responsive market approach (Table 56).

Table 56: Factor 9: Market approach and marketing

Theory	Manufacturing practice	Building practice
• Firms in an integrated supply chain are engaged in a joint market approach, led by the focal form, via collaborative marketing strategies, e.g. joint venture, marketing channelling or co-development. • The main aims of such approaches are increased responsiveness and anticipation of markets, as well as decreased time to market and dependence on variability of demand. • The concept of postponement has been a major market strategy to meet such aims. • In the context of building, supply chain integration has particularly been mentioned as an opportunity to reduce costs and therefore prices, as well as lead times by improving logistics, thereby improving revenues and increasing 'value for money' for clients. • Postponement in building has been pursued since the 1960s as in the open building movement and with subsequent followers promoting so-called industrialisation of building.	• Focal firms have organised the effective and efficient delivery of products through integrated sales channels, while also organising effective and efficient information capture of customer demand and market information, e.g. new trends to anticipate. • All efforts are aimed at responsive order fulfilment, e.g. via direct sales to customers. Forecasting of market developments is fed into the supply chain process. The aim is reduction of time to market and price. Products are assembled or engineered to order and adjusted to regions. Future technology is anticipated and installed in new products. • In retailing, it is important to maintain customer relations, releasing new products constantly for an exclusive and up-to-date image.	• Focal firms have only managed to integrate the sales channel in the case of being able to step across the intermediate stages or parties that traditionally stand between themselves and the user. • Some firms have chosen to be their own client and to deliver full service accommodation to a user. These products are generally aimed at niche markets in building. • Other market approaches include marketing and sales of integrated product directly from the factory, repetitive sales to large clients by offering a discount on a subsequent project, or marketing the product via other firms as well, such as the architect.
Relationship		**Leads to**
The integration of the sales channel to the end market.		A closed loop of market push towards the market while capturing market information and new trends from the market.
Process implications		**Product implications**
The processes and particularly the sales process need to be organised for speed and responsiveness.		Reconfigurability and marketability of products need to be at a high level.
Means		**Goals**
• Increase efficiency of the sales channel. • Improve market responsiveness.		• Increased product success. • Improved competitiveness.

Factor 10: Cultural alignment

Firms in an integrated supply chain have to align their respective business cultures and approach the supply chain as 'one firm'. Increased collectivism nurtures long-term relations in the supply chain (Table 57).

Table 57: Factor 10: Cultural alignment

Theory	Manufacturing practice	Building practice
• Firms will align their respective business cultures and virtually simulate the supply chain being 'one firm'. Firms engaged in an integrated supply chain foster cross-cultural understanding and create a culture of collaboration throughout the supply chain and in the broader network of firms. • The success of cultural alignment in the supply chain depends on the power balance and trust between firms. Generally, powerful focal firms who are able to build trust are more likely to achieve higher levels of cultural alignment within their supply chain. • A collectivist culture facilitates the development of understanding among the firms of a supply chain. Collectivism nurtures long-term relations. • The culture within building is a typical project culture and relatively informal compared to the more formal corporate culture of manufacturing.	• Collective cultures have been observed to be nurtured by the focal firm and used to improve the joint performance and collective competitiveness. • The culture of togetherness and being virtually one firm has been implanted at all the firms in the supply chain and maintained by formalised supplier development programmes and supplier events. • Supplier networks are approached as families imposing the corporate standards onto supply chain partners. Suppliers will need to subscribe to the supply chain codes. In addition, the focal firm makes an effort aimed at building trust and achieving transparency.	• In a few cases, supplier development and family approaches have been observed, and the management of the focal firm adopted their suppliers, but in a far less formalised and less far reaching way than in the manufacturing cases. • In other cases, clients have taken the lead involving contractors, subcontractors, and even residents in a joint supply chain movement. • The relatively weak corporate cultures and the lack of shared values among firms and individuals within the building supply chain have been mentioned as reasons for the low performance of building.
Relationship		**Leads to**
A balanced approach of the focal firm to power, trust and shared values, and imposing this in a constructive manner on the firms along the supply chain,		Increased commitment and collectivism among those firms vis-à-vis the focal firm.
Process implications		**Product implications**
Power and trust need to be balanced in all processes to align the cultures through all processes.		The end product needs to be embraced by all parties as their own.
Means		**Goals**
• All firms are culturally aligned by the focal firm. • High levels of commitment to the supply chain.		• Increase collectivism of the supply chain as 'one firm'. • Improved exposure and the image of the supply chain as a collective whole.

Factor 11: Human resource management

Firms will need to install a joint approach to personnel, such as forming joint teams across the extended business processes of the firms involved in the integrated supply chain. Team building and integration are essential to be able to unlock the potential advantage of the joint teams (Table 58).

Table 58: Factor 11: Human resource management

Theory	Manufacturing practice	Building practice
• As a distinct part of cross-corporate cultural alignment, firms have installed a joint approach to personnel, e.g. forming joint teams along the extended business processes of the firms involved. • Team building and integration have shown to be essential parameters to be able to unlock the potential advantage of the joint teams. • Building thrives on team performance. However, the loose personnel connections per project and the weak production system of building mitigate this effect and create a central role for the project manager, with therefore a high risk of failure.	• Focal firms attach much value to the level and quality of personnel contact with suppliers as a basis for maintaining long-term partnerships, e.g. via joint teams, personnel exchange and guest engineers. • The aim is to foster the notion of 'one firm' via intensified interaction and work exchange between individuals. For instance, the continual presence of personnel of the focal firm at the suppliers' plants is seen as a key connection in the supply chain.	• Intensified personnel exchange is generally limited to project teams. In the cases of intensified personnel exchange, this has been aimed at joint staffing on the management and project levels in the case of joint legal partnerships, or on an engineering level of joint engineering teams with suppliers. • In a few cases, a joint workforce on site with subcontractors was introduced based on lean techniques.
Relationship		**Leads to**
A structural approach to personnel and knowledge exchange, and joint staffing between firms in the supply chain.		Fostering the togetherness and the functioning of the supply chain as a collective whole.
Process implications		**Product implications**
Joint operation and management of processes.		Joint responsibility of products delivered.
Means		**Goals**
• Joint staffing and personnel exchange. • Fostering collective action by firms and individuals.		• Increased productivity and knowledge level of individuals. • Improved team performance of firms and individuals.

9.1.2 Relating and clustering the theoretical factors

The theoretical factors imply relations between measures contributing to goals, as identified in the tables above. Those relations will be used to categorise the factors by their goals (Table 59). In order to do so, ten goals have been identified and condensed from the collection of goals in the tables above. The relations between the *eleven factors* and the *ten goals* will next be the basis for reducing both the factors and the goals. Factors contributing to the same goals, and vice versa, goals related to the same factors, will be further analysed to find *patterns* leading to the shaping of the hypotheses further on in this chapter.

Table 59: Categorising the factors by relations to goals

Factors	Relationship leading to goals	1. improved flow of production	2. Increased productivity, joint knowledge level	3. Improved market responsiveness, time to market	4. improved joint competitiveness of chain	5. Improved competitiveness of the focal firm	6. Increased commercial success of products	7. Improved manufacturability of products	8. Improved total quality of products	9. Increase collectivism, image of supply chain	10. Improved team performance of firms, individuals
1. Repetitiveness	Coherent approach to increased repetitiveness of products and processes	X	X					X			
2. Integration of business activities	Integration of business activities by focal firm through internalisation and increased decentralised control					X	X				
3. Partner sourcing and collaboration strategies	Establishment of strategic, long-term collaboration by focal firm with the supply chain		X		X					X	
4. Integration of operations and processes	Seamless integration of operations and processes through the supply chain	X	X								
5. Planning and logistics control	Integrated planning and logistics across the entire supply chain	X		X							
6. Quality management	Application of total quality management across all corporate boundaries in the supply chain								X		
7. Information exchange	Commitment of firms to information flow through the supply chain	X		X	X						
8. Product development and design	Product development based on integrated contribution of all functions in the supply chain						X	X	X		
9. Market approach and marketing	Integration of the sales channel to the end market			X	X	X	X				
10. Cultural alignment	Balanced approach of focal firm to power, trust, shared values, imposing this on the supply chain									X	
11. Human resource management	Structural approach to personnel and knowledge exchange, joint staffing between supply chain firms		X							X	X

Table 60: Relating the factors

Factors	Relationship leading to goals	1.Repetitiveness	2.Integration of business activities	3. Partner sourcing and collaboration strategies	4. Integration of operations and processes	5. Planning and logistics control	6. Quality management	7. Information exchange	8. Product development and design	9. Market approach and marketing	10. Cultural alignment	11. Human resource management
1. Repetitiveness	Coherent approach to increased repetitiveness of products and processes	X		X	X	X		X	X			X
2. Integration of business activities	Integration of business activities by focal firm through internalisation and increased decentralised control		X						X	X		
3. Partner sourcing and collaboration strategies	Establishment of strategic, long-term collaboration by focal firm with the supply chain	X		X	X			X		X	X	X
4. Integration of operations and processes	Seamless integration of operations and processes through the supply chain	X		X	X	X		X				X
5. Planning and logistics control	Integrated planning and logistics across the entire supply chain	X			X	X		X		X		
6. Quality management	Application of total quality management across all corporate boundaries in the supply chain						X		X			
7. Information exchange	Commitment of firms to information flow through the supply chain	X		X	X	X		X		X		
8. Product development and design	Product development based on integrated contribution of all functions in the supply chain	X	X				X		X	X		
9. Market approach and marketing	Integration of the sales channel to the end market		X	X		X		X	X	X		
10. Cultural alignment	Balanced approach of focal firm to power, trust, shared values, imposing this on the supply chain			X							X	X
11. Human resource management	Structural approach to personnel and knowledge exchange, joint staffing between supply chain firms	X		X	X						X	X

Based on the categorisation of the factors related to their goals, we will now relate and find *mutual connections* between the *factors* (Table 60). Apparently, certain factors such as repetitiveness have more connections to other factors than factors such as quality management do. However, in this analysis only *direct connections* have been identified, both in the literature and in the empirical findings. In addition, the factors have not been weighted, nor have their relations to the goals.

Based on the correlations between the factors as identified above, we will now be able to group close connections together, leading to the identification of clusters of *related factors* (Table 61). The clustering of the factors implies a rearrangement of factors that are apparently interrelated, pointing in the direction of the hypotheses to be formulated further in this chapter. The clustering is based on the schematic application of the mathematical approach of the *graph theory* (Harary & Norman 1953). The graph theory is based on the concept of graphs as abstract configurations of sets of elements and their mutual relations (De Ridder 1994). In this thesis, the application of the graph theory has led to the *decomposition* of the relations between factors and goals, followed by the *reconfiguration* of the relations into clusters with stronger *intrinsic coherence*. This has been applied below by *permutations* of the rows and columns of the table above of related factors into clusters of factors in the table below (Table 61). The process of permutation has been characterised by a visual and iterative procedure, shifting rows and columns and manually clustering related factors. As a result, four relatively independent *clusters of factors* have been identified and labelled as follows: integration, market development, collaboration, and quality.

9.1.3 Relating and clustering the goals

Having identified the clusters of factors, we will now identify the *clusters of goals* to be combined with the clusters of factors further in this chapter. Both the clusters of factors and goals, and the relations between both clusters will eventually be rephrased as hypotheses.

As a *first clustering* of the goals, we will sort the ten initial goals according to their correspondence to the four clusters of factors identified above . The first cluster of factors seems to have a rather strong and concentrated influence on the goals (Table 62). The goals mainly concentrate on *improved flow of production*. The second cluster of factors has a weaker and less concentrated influence on the goals. These goals particularly refer to the *improved commercial success* of products. The third cluster of factors has a decreasing influence on the goals. These goals mainly point to *improved collectivism and image* of the supply chain. To conclude, the remaining factor of quality management has a narrow influence on *improved total quality*. These four correlations between the four clusters of factors and the ten initial goals implicitly represent a first clustering of the goals. They point towards four main goals, but like the factors, the goals have not been weighted against each other.

Table 61: Clustering the factors

Factors	Relationship leading to goals	1. Repetitiveness	4. Integration of operations and processes	7. Information exchange	5. Planning and logistics control	8. Product development and design	2. Integration of business activities	9. Market approach and marketing	3. Partner sourcing and collaboration strategies	11. Human resource management	10. Cultural alignment	6. Quality management
1. Repetitiveness	Coherent approach to increased repetitiveness of products and processes	X	X	X	X	X			X	X		
4. Integration of operations and processes	Seamless integration of operations and processes through the supply chain	X	X	X	X				X	X		
7. Information exchange	Commitment of firms to information flow through the supply chain	X	X	X	X			X	X			
5. Planning and logistics control	Integrated planning and logistics across the entire supply chain	X	X	X	X			X				
8. Product development and design	Product development based on integrated contribution of all functions in the supply chain	X				X	X	X				X
2. Integration of business activities	Integration of business activities by focal firm through internalisation and more decentralised control					X	X	X				
9. Market approach and marketing	Integration of the sales channel to the end market			X	X	X	X	X	X			
3. Partner sourcing and collaboration strategies	Establishment of strategic, long-term collaboration by focal firm with the supply chain	X	X	X				X	X	X	X	
11. Human resource management	Structural approach to personnel and knowledge exchange, joint staffing between supply chain firms	X	X						X	X	X	
10. Cultural alignment	Balanced approach of focal firm to power, trust, shared values, imposing this on the supply chain								X	X	X	
6. Quality management	Application of total quality management across all corporate boundaries in the supply chain					X						X

Integration

Market development

Collaboration

Quality

Table 62: Identifying main goals based on the clusters of factors

Factors	Relationship leading to goals	1. Improved flow of production	2. Increased productivity, joint knowledge level	3. Improved market responsiveness, time to market	4. Improved joint competitiveness of chain	5. Improved competitiveness of the focal firm	6. Increased commercial success of products	7. Improved manufacturability of products	8. Improved total quality of products	9. Increased collectivism, image of supply chain	10. Improved team performance of firms, individuals
1. Repetitiveness	Coherent approach to increased repetitiveness of products and processes	X	X					X			
4. Integration of operations and processes	Seamless integration of operations and processes through the supply chain	X	X								
7. Information exchange	Commitment of firms to information flow through the supply chain	X		X	X						
5. Planning and logistics control	Integrated planning and logistics across the entire supply chain	X		X							
8. Product development and design	Product development based on integrated contribution of all functions in the supply chain						X	X			
2. Integration of business activities	Integration of business activities by focal firm through internalisation and increased decentralised control					X	X				
9. Market approach and marketing	Integration of the sales channel to the end market			X	X	X	X				
3. Partner sourcing and collaboration strategies	Establishment of strategic, long-term collaboration by focal firm with the supply chain		X		X					X	
11. Human resource management	Structural approach to personnel and knowledge exchange, joint staffing between supply chain firms		X							X	X
10. Cultural alignment	Balanced approach of focal firm to power, trust, shared values, imposing this on the supply chain									X	
6. Quality management	Application of total quality management across all corporate boundaries in the supply chain								X		

Table 63: Relating the goals

Goals		1. Improved flow of production	2. Increased productivity, joint knowledge level	3. Improved market responsiveness, time to market	4. Improved joint competitiveness of chain	5. Improved competitiveness of the focal firm	6. Increased commercial success of products	7. Improved manufacturability of products	8. Improved total quality of products	9. Increased collectivism, image of supply chain	10. Improved team performance of firms, individuals
1. Improved flow of production		X	X	X	X			X			
2. Increased productivity, joint knowledge level		X	X		X			X		X	X
3. Improved market responsiveness, time to market		X		X	X	X	X				
4. Improved joint competitiveness of chain		X	X	X	X	X	X			X	
5. Improved competitiveness of the focal firm				X	X	X	X				
6. Increased commercial success of products				X	X	X	X	X	X		
7. Improved manufacturability of products		X	X				X	X	X		
8. Improved total quality of products							X	X	X		
9. Increased collectivism, image of supply chain			X		X					X	X
10. Improved team performance of firms, individuals			X							X	X

As a *second clustering* of the goals, corresponding to the clustering of the factors, we will now relate and cluster the ten goals mutually, based on the previous categorisation of the initial eleven factors and the ten goals (Table 59). Following this categorisation, we will find *mutual connections* between the goals (Table 63). The connections between the goals seem to concentrate in *four clusters of goals*. These clusters do not necessarily correspond to the four clusters of factors. We will analyse this further on in this chapter as part of the formulation of the hypotheses.

Table 64: Clustering the goals

Goals		1. Improved flow of production	2. Increased productivity, joint knowledge level	4. Improved joint competitiveness of chain	3. Improved market responsiveness, time to market	5. Improved competitiveness of the focal firm	6. Increased commercial success of products	7. Improved manufacturability of products	8. Improved total quality of products	9. Increased collectivism, image of supply chain	10. Improved team performance of firms, individuals
1. Improved flow of production		X	X	X	X			X			
2. Increased productivity, joint knowledge level		X	X	X		Productivity		X		X	X
4. Improved joint competitiveness of chain		X	X	X	X	X	X			X	
3. Improved market responsiveness, time to market		X		X	X	X	X	Market position			
5. Improved competitiveness of the focal firm				X	X	X	X				
6. Increased commercial success of products				X	X	X	X	X	X		
7. Improved manufacturability of products		X	X				X	X	X	Product	
8. Improved total quality of products							X	X	X		
9. Increased collectivism, image of supply chain			X	X			Team performance			X	X
10. Improved team performance of firms, individuals			X							X	X

In the same way that the factors have been clustered above by application of the graph theory, the goals presented in the previous table will also be rearranged and clustered by *permutations* of rows and columns into clusters of goals in the following table (Table 64). Apparently, just one permutation leads to the identification of four relatively independent *clusters of goals*, labelled as follows: productivity, market position, product, and team performance. Those four clusters of goals appear to correspond to the main goals previously identified above (Table 62), respectively:

1. Productivity → improved flow of production
2. Market position → improved commercial success
3. Product → improved total quality
4. Team performance → improved collectivism and image

9.1.4 Shaping the hypotheses

Following the above clustering of factors and goals, the *hypotheses* resulting from this thesis can now be shaped. Based on the clusters of *factors*, the *relations* can first be induced between the respective *actions* embedded in the factors, the *focus* of the actions and their supposed *effects* (Table 65). The labels of both the clusters of factors and the clusters of goals follow from the above and represent collections of factors and goals rather than individual factors and goals. Moreover, the factors and goals refer to supply chains rather than factors and goals of individual firms.

Table 65: Clusters of factors

Cluster of factors	Actions	Focus	Effects
1. Integration	Increasing the repetitiveness, integration and commitment in the supply chain	Integrated products, processes, flow of production and information, planning and logistics	• Improved flow of production, information • Increased productivity • Improved effectiveness • Improved efficiency
2. Market development	Improving the connection to the end market, and the control of marketing activity by the focal firm	Integrated contributions to design and product development, and alignment of business activities and sales channel to the end market	• Improved market responsiveness • Improved joint competitiveness • Improved competitiveness of focal firm • Increased commercial success of products
3. Collaboration	Structuring and aligning the individual activities and contributions of firms and individuals	Sourcing and collaboration, power and trust, shared values, joint staffing and knowledge exchange	• Increased collectivism • Improved team performance of firms • Improved team performance of individuals
4. Quality	Managing the total quality of products and processes	Objective and subjective quality in all quality systems	• Improved and continuous total quality delivered through the supply chain

Although the clusters of *factors* cannot be ordered based on their different importance or hierarchy in the clusters, a certain *logical order* can be distinguished implicitly based on a reflection of *causal relations*. Increased collaboration and alignment (3) would lead to increased integration and commitment (1). This would lead to improved quality (4), thus supporting market development and improved marketing activity in the supply chain (2).

Next, based on the clusters of *goals*, the *relations* can be induced between the higher *aims* of those clusters of goals, their *background*, and the *actions* contributing to them (Table 66). For the purpose of correlation of the clusters of factors with the clusters of goals, and the formulation of the hypotheses, the third and the fourth clusters of goals have been swapped.

Table 66: Clusters of goals

Cluster of goals	Aim	Background	Actions
1. Productivity	Increased efficiency and effectiveness of the supply chain	Effective and efficient delivery of end products through the supply chain to the end market	• Increasing flow through the supply chain • Increasing integration and repetitiveness of production activities • Increasing commitment and alignment to the production system of the supply chain
2. Market position	Strong connection and position to the end market	Improved competitive position of the supply chain and the focal firm vis-à-vis other supply chains and focal firms	• Increasing market control by the focal firm • Mobilising the supply chain for increased joint and focal firm competitiveness • Involving the supply chain and sales channel for increased commercial success of products
3. Team performance	Firms and individuals in the supply chain operating as an integrated team	Viewing the supply chain as one virtual organisation, i.e. an extended enterprise	• Promoting collective action and alignment • Promoting commitment to the supply chain • Shifting focal firm power and trust of firms in the supply chain
4. Product	Unconditional delivery and quality of integrated products to the end customer	Total quality and improved manufacturability of products is essential for improved effectiveness and efficiency	• Promoting commitment of all firms to total quality of products and processes • Establishing continuous improvement measures and quality culture

Like the clusters of factors, the clusters of *goals* also cannot be ordered based on different importance or hierarchy. Again, however, a certain *logical order* can be distinguished implicitly based on a reflection of *causal relations*. Improved team performance (3) would lead to increased productivity (1). This would also trigger improved product delivery (4), and as a result, improve the market position of the supply chain (2).

Based on the above relations induced from the clusters of factors and the clusters of goals, the *four hypotheses* resulting from this thesis are now at last formulated.

Table 67: Formulation of hypotheses, connections to theory and manufacturing, and applicability to building.

Relation	Hypotheses	Connection to existing theory and manufacturing practice	Applicability to building practice
1. Integration → Productivity	Integration of the extended delivery process through the supply chain leads to increased productivity of the firms involved.	Organisational and production theories and corresponding concepts in manufacturing practice often imply integration of processes leading to operational improvements including increased productivity.	The project culture in building has fostered a fragmented and temporary production organisation. Supply chain strategies must be able to flexibly promote higher levels of integration among a relatively large number of dispersed and often small firms.
2. Market development → Market position	Improving the connection of the integrated supply chain to the end market leads to improved competitiveness of the supply chain as a whole and the focal firm in particular.	Economic theory and marketing concepts, particularly in retail sectors, propagate increased connectedness and responsiveness to the end market to improve competitiveness primarily for the focal firm, i.e. manufacturer or retailer, and to a lesser degree for the supply chain as well.	The disconnection of the focal firm to the end market, and the many intermediate parties between the focal firm and the end customer imply forging a delegated market connection. Gaining market position is a collective matter instead of an individual aim of the focal firm.
3. Collaboration → Team performance	Aligning the contributions of firms and individuals to the supply chain improves the joint performance as well as the individual performances.	Network theories as well as inter-firm and inter-human concepts point to increased alignment of individual contributions of firms and humans involved in grouped operations leading to improved operational performance. This is a leading principle in manufacturing practice, and generally quite effectively and formally organised between firms.	Aligning team performance in building is generally associated with site production to overcome the high levels of fragmentation of the many specialist firms and suppliers organised informally. This idea must be extended to also include contributions in the development and design. More effective and formal approaches to cross-corporate team formation are needed.
4. Quality → Product	Implementing total quality management to the integrated delivery process through the supply chain leads to delivery of products with unconditional quality to the end customer.	The concept of total quality management has been an integrated part of production theory and standing practice in manufacturing industries for decades and has been a key contributor to improved product delivery by firms and the idea of zero defects in supply chains.	In particular, the traditional disconnection between design and production needs to be addressed, e.g. design quality versus technical quality. This implies integrated quality management, and integrating design and production to improve buildability and the commitment of firms to implementing total quality.

The four hypotheses respectively reflect *four relations* between the reduced clusters of factors on the one hand and the reduced clusters of goals on the other. The hypotheses show connections with existing theory and supply chain integration applications in manufacturing, and suggest that the building industry could *change* in this direction as well. Reversely, for improved applicability and operationalisation of the hypotheses to the context of building, their *adaptation* to building would be helpful, specifically addressing the industry's unique combination of characteristics (Table 67).

The *first hypothesis* implies a relationship between the increased integration and the increased productivity of the supply chain. The relationship suggests that higher levels of integration of the extended delivery process through the supply chain, including the connections between respective business processes of firms involved, leads to increased effectiveness and efficiency throughout the supply chain. The *second hypothesis* implies a relationship between the improved connection to the end market and the improved market position of the supply chain. The relationship suggest that the improvement of the connection and responsiveness of the integrated supply chain to the end market, supported by increased control of collective market activities and the integrated sales channel towards the end market by the focal firm, leads to improved commercial success of products delivered to the end market, and higher competitiveness of the supply chain and the focal firm in particular as a result. The *third hypothesis* implies a relationship between increased collaboration and improved team performance. The relationship suggests that the intensified alignment of the contributions of firms and individuals to the supply chain, and structuring their collective activities as an integrated team, improves the joint performance as well as the individual performances. The *fourth hypothesis* implies a relationship between total quality management and unconditional product delivery. The relationship suggest that the implementing of total quality management to the integrated delivery process throughout the supply chain, including a commitment of firms to continuous improvement of quality of components, manufacturability of integrated products and quality of processes, leads to the delivery of end products with unconditional quality to the end customer.

Reflection on the hypotheses and theoretical contribution

These hypotheses must be regarded as the closure of the theory building process and therefore the resulting theoretical contribution of this thesis, to be researched and tested in further research. The theoretical contribution must be interpreted as a proposed *adaptation or extension* of *existing theory* rather than the development of a new theory. The theoretical extension represents the concept of supply chain integration for the building industry at which this thesis has been aimed, however, the hypotheses have been put in *general terms* rather than in building-specific terms. Based on the formulation of the hypotheses, one could argue that the concept of supply chain integration in the building industry, indeed, reflects an adaptation or extension of existing theory for improved *application* to the building supply chain and to the context of building in general. Alternatively, the hypotheses imply a *shift of building practice*, applying more comprehensive supply chain integration strategies following existing theory and the example of manufacturing. Finally, although the central topic of investigation has been supply chain integration, *additional explanations* for the phenomena encountered and found in the thesis could also be furnished by competing or related concepts that have developed in parallel in building theory and practice, e.g. partnering and lean construction. This might be a subject of further research to extend or test the hypotheses.

9.2 Answers to research questions

The hypotheses presented have implicitly *answered the central research question*: How would a concept of supply chain integration for the building industry look, based on constructs from existing theory, examples of advanced applications of supply chain integration in manufacturing, and initial developments towards supply chain integration in building?

This central research question as well as the six partial research questions formulated in Chapter 1 have functioned as 'process questions' leading to the formulation of the hypotheses above, and have been answered through the course of this thesis, rather than being the answers the research had been aimed at ultimately.

1 What constitutes the concept of supply chain integration in current theory?
In construction management research, various theoretical indications have been found pointing towards supply chain integration. In general theory, *four domains of theory* have been identified related to supply chain integration: economic, production, organisation and social theory. The theoretical framework of the thesis has drawn elements from those four domains. Further operationalisation of the theoretical framework has led to an analytical model of supply chain integration containing the following *factors of analysis*: integration of business activities, collaboration and sourcing strategies, integration of processes and operation, planning and logistics, information sharing, quality management, product development and design, market approach and marketing, cultural alignment, and human resource management. The theoretical findings have provided initial grounding for the development of the concept of supply chain integration for the building industry in the synthesising part of this thesis (Chapters 3 to 5).

2 What supply chain integration practices have been applied in manufacturing?
The thesis has found applications of supply chain integration in *six sectors of manufacturing*: automotive, aerospace, computers, electronics, clothing/textile and food/grocery. The applications have been explored and described following the factors of analysis of the analytical model of supply chain integration developed in this thesis. The exploration of the manufacturing cases has led to insights and explana-

tions of how supply chain integration has been applied in the respective cases. In all cases, the focal firm has shown considerable *dominance* of the supply chain and an urge to excel operationally with supply chain partners. Supply chain partners have been found to be highly *committed* and *aligned* to the focal firm and the supply chain. The main aims of this approach include achieving higher levels of *collective competitiveness*, market responsiveness and market push with continually new products and new technology. The findings of the case studies in manufacturing have shown varied evidence of application of the theoretical elements of supply chain integration as found in the theoretical part of the thesis (Chapter 6).

3 What supply chain integration practices have been applied in building?

The thesis has explored applications of supply chain integration from *five positions* in the building supply chain: client organisations, project developers, designers, builders and suppliers. In accordance with the manufacturing cases, the applications in building have been explored and described following the factors of analysis of the analytical model of supply chain integration. The applications by the firms studied have shown considerable *differences* in their approach and advancement. The preferences per firm type for particular subjects varied based on the position, type of business and power of the particular firm in the supply chain. Most cases, however, had shown an inclination towards increased levels of integration of business functions while other supply chain activities were left relatively untouched, such as integrated information and quality management. The findings of the building case studies have demonstrated relatively less advanced, less comprehensive, quite pragmatic and rather informal approaches to supply chain integration, as compared to theory and manufacturing (Chapter 7).

4 How do applications of supply chain integration in building versus manufacturing relate?

The within- and cross-case analyses of both applications in manufacturing and in building have led to further insights, pattern matching and explanation building within and between both groups of applications. Particularly the cross-case analysis between both groups has led to the identification of distinctive similarities and differences in the application of supply chain integration. *Similarities* have mainly been found in the reliance on long-term relations with external supply chain partners, joint planning and logistics, and installation of joint teams between firms. However, the level and scope of these issues were *more restricted in building* as compared to manufacturing. Particular issues of supply chain integration of major importance in manufacturing have shown to be addressed to a lesser degree in building, such as integrated market approaches, product development and cultural alignment. Typically underrepresented in building were integrated approaches to information management and quality management, which are regarded as cornerstones of supply chain integration both in theory and in manufacturing paradoxically (Chapter 8).

5 How do the applications of supply chain integrations in practice relate to theory?

Theory and practice have been related in terms of eleven factors of supply chain integration and their goals to improve the functioning of the supply chain. The *factors* include increased repetitiveness of products, processes and organisational formats, increased integration and control of business activities, entering into strategic and long-term arrangements, integration of operations and processes, integrated control of planning and logistics, introduction of total quality management, integrated information and knowledge sharing, integrated product development and design, close connections with the end market, alignment of cultures along the supply chain, and personnel exchange and joint staffing between firms within the supply chain. Related to the theory, on all aspects of all factors, applications of supply chain integration in *manufacturing* tend to be present and implemented more intensively, broadly and dominantly by the *focal firm* throughout the supply chain than in building (Chapter 9).

6 What constitutes a concept of supply chain integration in building?

The confrontation of the collective empirical findings of the thesis with theory has led to the generation of *four hypotheses* of supply chain integration in building. The hypotheses represent partial descriptions of what constitutes a concept of supply chain integration in building, and collectively form the theoretical contribution as it emerged from this thesis. In general, this *emergent theory* would imply a redefinition of connections between firms in the building supply chain towards higher levels of repetitiveness and integration of products, business processes and inter-firm relations among the different firm types from clients to suppliers, and the supply chain evolving towards an *extended enterprise* or quasi-firm as it were a single firm persisting to exist, beyond the scope of separate projects (Chapter 9).

9.3 Contributions to theory and method

Contribution to theory

This thesis has contributed a theoretical concept or an emergent theory of supply chain integration in building. The emergent theory suggests a *new way of looking* at the organisation and coordination of processes and relations across corporate boundaries in the building supply chain. As part of this, the thesis has suggested *new combinations of theoretical concepts* to the context of building:

- It proposes the integration of business activities and processes along the supply chain, and applying total quality management and integrated control of planning and logistics.
- It suggests strategic cooperation and intensified information and knowledge exchange, and joint benefit sharing among firms in the supply chain.
- It leads to joint and proactive product development and responsive market ap-

proaches to increase collective profitability and competitiveness as a deliberate aim.
- It fosters a collaborative culture of togetherness and collectivism among firms and people, and more effective and efficient working as a result.

Because of the tendency of the thesis to focus on *housing*, the concept developed should therefore particularly apply to housing. However, since it originated from general theory initially, the concept may as well be applicable to *other sectors* within building such as commercial building or civil construction, or even to sectors *beyond the realm of building*, if the concept were extended and possibly modified to those contexts. The consequences of such theoretical extension, however, will require a further debate about the *contextual differences* between sectors of industry, for instance, with regard to the possibilities and the achievable level of repetition, integration and centralisation in those sectors. Further research would be needed to explore and develop those paths of theoretical extension.

Contribution to method

This thesis has contributed methodological novelty by applying *Eisenhardt's* theory building approach using case study research to a subject previously not approached with such a research method. Although the method has often been applied in generic management research, this had previously not been the case in the domain of construction management research and not to the full extent of the method. The thesis has thus led to fresh insights using this method, adding to new theory and thus producing new knowledge for supply chain integration in building. In essence, the thesis has aimed to contribute to research practice by the application of this fairly *novel approach* and as a consequence contributed new methodological experience to the field of construction management research in particular. Further development of this approach in construction management research is encouraged.

9.4 Implications for building practice

Extended enterprise as a governance mechanism

The extended enterprise has been proposed as a governance mechanism for the integrated building supply chain. The effect would be that the management and control of the supply chain will become more *centralised* into an integrated structure and single *hierarchy*, factually or virtually. This type of governance of activities makes it possible for integrated chains of firms to present a total product to the end market and the end user. However, the current competition for *control* between firms within the building supply chain, unbalanced power, lack of trust between firms and the loss of *autonomy* will probably form obstacles to achieving *integrated governance* of building supply chains. Supply chain integration would therefore need to consider dealing with these obstacles by ways of achieving multi-project and multi-organisational arrangements in the building supply chain. Client organisations could have a lead role in this process and be the 'reseller' of the end product to the end customer, for instance, by involving a contractor to be the 'manufacturer' of the integrated product. However, any firm in the building supply chain could adopt a role of *focal firm* and internalise integrated business activities or either involve supply chain partners. Either way, when governed as an extended enterprise, the building supply chain will achieve a higher level of integration and repetitiveness, and centralised control of the network of firms, resembling the situation in manufacturing.

Centralised control for improved joint performance

Centrally coordinated building supply chains, however, in contrast to the general situation in manufacturing, must not merely be directed towards defined operational optimisation, but also towards being able to flexibly enhance the *transfer* of expertise and systematic *feedback* between parties, and ultimately towards joint *value maximisation*. Supply chain integration in building will need to deal with a *diversity of demand* and reduce the resulting unpredictability. It will need to efficiently integrate diverse inputs and new technologies, and harnesses the delivery process against the impact of *'environmental turbulence'* operative in building. The emergence of centralised forms of organising supply chains would make it possible to present a total product with quality guarantees to the market. Integrated delivery of total products will likely introduce such roles as the integrated *service provider* and the *system integrator* to the building supply chain, much like the manufacturer of the integrated product in manufacturing. Those roles can be combined in one *focal firm* or could be shared among a number of *allied firms*, depending on such issues as the power balance between firms, or for pragmatic reasons such as the division of responsibilities between firms.

Focal firm as a leader in the supply chain

The role of the system integrator could be one central role or could alternatively be divided into two partial roles: the demand system integrator and the supply system integrator. For instance, a client organisation could take up the *demand integrator* role and the main contractor could take up the supply integrator role. In many cases, the client organisation represents many users and other stakeholders, often referred to as the 'clients behind the client'. On their behalf, the client organisation takes care of the procurement of a built facility. The way in which procurement takes place influences the degree of supply chain integration, which again influences the supply chain performance and ultimately the value that is delivered to the client. Through adoption of new procurement strategies and methods, the client can facilitate supply chain integration. On the *supply side* as well, independently from the demand, parties can evolve towards more integrated production and business strategies, through project-independent collaboration with other firms in the supply chain as well as internalisation of business or production activities within their own firms. In both cases, operational and competitive advantages through greater productivity and efficiency and delivering better value to the market must be the drivers for this kind of supply chain integration. Normally, this development from the supply side is led by one focal firm or a complementary group of firms , whether it be a developer, architect, builder or a group of suppliers.

The building case studies in this thesis have discovered various applications of supply chain integration in building practice, echoing the above implications, depending on the firm type that represented the focal firm. Extrapolating from the descriptions of the building case studies, the concept of supply chain integration in the building industry could be *specified practically* for each of the five firm types in somewhat more prescriptive terms (Table 68).

9.5 Directions for further research

The research presented in this thesis represents, as all research does, a temporary instance in a continuum of research efforts. Typically, hypothesis generating research marks many directions for further research to test the hypotheses postulated and further develop theory. Based on this thesis, further research could be conducted in these directions, amongst others, aimed at the issues of further theoretical development, methodological advancement and practical applicability:

- Theory
  - Further development and testing is needed of the hypotheses that emerged from this thesis. Specifically, *internal* relationships and validity of the emergent theory will need to be strengthened.
  - Generalisation of the emergent theory is needed by means of empirical evidence confirming or disconfirming the theory, and relating the emergent theory further to general theory. This may include additional theoretical domains that this thesis had not yet mobilised, e.g. financial theory. This will improve *external* relationships and validity of the emergent theory.
- Method
  - Further application and advancement of *Eisenhardt's* theory building method by using case study research in construction management research is encouraged. The method has demonstrated to be useful in the development of an organisational concept, as used in this thesis. The method may well be useful for other types of concept development or theory building.
  - In addition to Eisenhardt's theory building method, *other research methods* that are well accepted in general management research could be used to further develop and test the emergent theory for supply chain integration in building developed in this thesis. Eisenhardt's method can be put in the tradition of *synthetic sense making* strategies for emerging theories (Langley 1999). Those strategies are aimed at producing relatively simple theoretical constructs that are rather general because they make sense of data from a number of cases. While theory development is a synthetic process,

Table 68: Specification of the concept of supply chain integration per firm type

	Clients	Project developers	Designers	Builders	Suppliers
1.Increased repetitiveness	Clients take a multi-project approach to procurement.	Developers develop product concepts used for multiple projects.	Designers develop concept design solutions for more projects.	Builders establish integrated process from design to delivery.	Suppliers predevelop integrated prefab subassemblies.
2. Increased integration and control of business activities	Clients concentrate on servicing users, and managing the governmental context.	Developers take concept design on board as a business activity.	Designers connect their design activities to development and building.	Builders include development, sales and design by delivering integrated solutions.	Suppliers connect to the technical engineering and site installation with the builder.
3. Strategic and long-term partnering arrangements	Clients enter into strategic arrangements with the supply chain and governments.	Developers enter into strategic arrangements with designers and builders.	Designers enter into product development with suppliers.	Builders enter into fixed supply relations with suppliers.	Suppliers establish mutual relations with other suppliers.
4. Integrated operations and processes	Clients deliver the building and manage entire development.	Developers take overall control of sales based on concept design.	Designers integrate product designing and engineering.	Builders integrate operations and processes with suppliers to flow.	Suppliers integrate their processes with the builder.
5. Integrated planning and logistics control	Clients plan operations and logistics based on development, involving builders.	Developers plan operations and logistics based on development, involving builders.	Designers plan operations and logistics based on design, involving builders.	Builders put in place advanced and integrated planning and logistics, e.g. lean.	Suppliers connect to and integrate with the planning system of the builder.
6. Total quality management	Clients prescribe total quality approach and specify proper compliance for supply chain.	From the project development, all quality standards are set for the entire supply chain.	Designers adopt quality standards in their design.	Builders integrate all necessary quality controls in their integrated process.	Suppliers assure the quality of subassemblies, connected to the quality system of the builder.
7. Integrated information flow and communication	Clients forward demand and usage information to the supply chain, and control communications.	Developers capture demand information, put in place integrated information and knowledge system for supply chain.	Designers connect and add to information system, fusing design and engineering in BIM.	Builders connect and add to information system and BIM, adding further technical/production information.	Suppliers connect and add to information system and BIM, adding further technical/product information.

8. Integrated product development and design	Clients deliver the building and manage development and design.	Developers develop and deliver integrated products as life cycle solutions.	Designers pool joint product development and design, involving supply chain.	Builders add integrated technical solutions to the design with suppliers.	Suppliers develop and deliver integrated subassemblies to integrated solution.
9. Integrated market channel closely connected to end market	Clients closely connected to market and users, translating information to development.	Developers capture market and client information and lead this into supply chain.	Designers anticipate market demand and trends with integrated concept designs.	Builders anticipate market demand with integrated product solutions, and add features to future products.	Suppliers anticipate demand with integrated subassemblies and develop future technology.
10. Aligned cultures along the supply chain	Clients start and promote the alignment of cultures in the supply chain.	Developers connect to the alignment and further align with designers and builders.	Designers connect to the alignment and further align with builders.	Builders connect to the alignment and further align with suppliers.	Suppliers connect to the alignment particularly with builders.
11. Personnel exchange and joint staffing	Clients install joint teams with supply chain partners, e.g. for strategic project planning.	Developers mobilise concept product teams with designers and builders.	Designers mobilise design teams with builders and suppliers.	Builders integrate teams and exchange personnel with suppliers.	Suppliers integrate personnel in joint teams with builder.

application of this kind of method could further develop the emergent theory, and enrich the methodological landscape of construction management research.

- Application
    - To expand on the relatively predominant focus on housing in this thesis, the emergent theory could specifically be tried and applied to commercial building, civil construction and *other segments of building* in order to develop a more generally valid and more widely applicable theory. The applications of supply chain integration in those other segments of building may vary considerably and as a result add new insights and elements to the understanding of supply chain integration in the wider building industry.
    - Research could be undertaken into the fairly complex structure of the *demand chain* in building, and the diversely regulated context of building. This includes the roles of politics, government, regulation and policies of competition, which in addition, vary considerably depending on each segment of building. Increased understanding of those *contextual factors* will increase insights into the conditions for and obstacles to the applicability of supply chain integration in building.

References

Abdullah, A.M. and Vicridge, I.G. (1999). "Best practice for multi-project management in the construction industry". In: *Proceedings COBRA*. RICS Foundation, 169-179.

Adler, P.S., Goldoftas, B. and Levine, D.I. (1999). "Flexibility versus efficiency?: a case study of model changeovers in the Toyota production system". *Organization Science,* 10 (1), 43-68.

AFN (2009). "Hooguit 1,5% faalkosten: integraal werkend architect heeft oog voor vakkennis van gespecialiseerde aannemers". Aannemersfederatie Nederland, *Bouwbelang,* 2 (6), 20-23.

Akkermans, H. (2001). "Emergent supply networks: system dynamics simulation of adaptive supply agents". In: *Proceedings 34th Hawaii International Conference on System Sciences*. 1-11.

Albino, V., Carbonara, N. and Giannoccaro, I. (2007). "Supply chain cooperation in industrial districts: a simulation analysis". *European Journal of Operational Research,* 177 (1), 261-280.

Amaral, D.C., Rozenfeld, H. and De Toledo, J.C. (2002). "Early supplier involvement: a review and proposal for new directions in research". *Brazilian Journal of Management of Product Development,* 2 (2), 25 p.

Amaratunga, D. and Baldry, D. (2001). "Case study methodology as a means of theory building: performance measurement in facilities management organisations". *Work Study,* 50 (3), 95-104.

Applebaum, H.A. (1982). "Construction management: traditional versus bureaucratic methods". *Anthropological Quarterly,* 55 (4), 224-234.

Armistead, C. and Mapes, J. (1993). "The impact of supply chain integration on operating performance". *Logistics Information Management,* 6 (4), 9-14.

Arrow, K.J. (1974). *The limits of organisation.* Norton, New York.

Asplund, E. and Danielson, U. (1991). *Räta ut byggsvängen.* In Swedish: Straightening the Building Roundabout. SBUF, Stockholm.

Axelrod, R. (1984). *The evolution of cooperation.* Basic Books, New York.

Baiden, B.K., Price, A.D.F. and Dainty, A.R.J. (2006). "The extent of team integration within construction projects". *International Journal of Project Management,* 24, 13-23.

Bair, J. and Gereffi, G. (2001). "Local clusters in global chains: the causes and consequences of export dynamism in Torreon's blue jeans industry". *World Development,* 29 (11), 1885-1903.

Baldwin, C.Y. and Clark, K.B. (2000). *Design rules.* MIT, Cambridge, MA.

Ballard, G. (1998). *Implementing Pull Strategies in the AEC Industry.* LCI White Paper - 1. Lean Construction Institute.

Ballard, G. (2005). "Construction: one type of project-based production system". In: *Proceedings SCRI Forum Event Lean Construction: The Next Generation.* 19 January 2005, SCRI, University of Salford, Salford. 14 p.

Ballard, G. and Howell, G.A. (1998). "What kind of production is construction?". In: *Proceedings 6th Annual Lean Construction Conference (IGLC-6).* 13 August 1998-15 August 1998, Guarujá, Brazil.

Ballard, H.G. (2000). *The last planner system of production control.* Doctorate thesis. University of Birmingham, Birmingham. 192 p.

Ballou, R.H. (1998). *Business logistics management.* Edition 4. Prentice Hall, London.

Barney, J.B. (1991). "Firm resources and sustained competitive advantage". *Journal of Management,* 17, 99-120.

Barratt, M. (2004). "Understanding the meaning of collaboration in the supply chain". *Supply Chain Management: An international Journal,* 9 (1), 30-42.

Barrett, P. (2000). "Systems and relationships for construction quality". *International Journal of Quality & Reliability Management,* 17 (4/5), 377-392.

Bechtel, C. and Jayaram, J. (1997). "Supply chain management: a strategic perspective". *International Journal of Logistics Management,* 8 (1), 15-34.

Beelaerts van Blokland, W.W.A. (2010). Value leverage by aerospace original equipment manufacturers. Delft University of Technology, Pijnacker. 220 p.

Beelaerts van Blokland, W.W.A., Santema, S.C. and Curran, R. (2010). "Lean supply chain management in aerospace." In: Blockley, R. and Shyy, W. (eds.). *Encyclopedia of Aerospace Engineering*, Volume 6. John Wiley & Sons, Hoboken.

Benheim, M. and Birchall, D. (1999). "Networks in the construction industry: emerging regulation procedures in the production process". In: *Proceedings Profitable Partnering in Construction Procurement: CIB W092 and TG23 Joint Symposium.* Thailand. Routledge, 63-75.

Bennett, J. and Peace, S. (2006). *Partnering in the construction industry: a code of practice for strategic collaborative working*. Butterworth-Heinemann, Oxford. 267 p.

Berry, D., Evans, G.N., Mason-Jones, R. and Towill, D.R. (1999). "The BPR SCOPE concept in leveraging improved supply chain performance". *Business Process Management Journal,* 5 (3), 254-274.

Berry, D., Towill, D.R. and Wadsley, N. (1994). "Supply chain management in the electronics products industry". *International Journal of Physical Distribution & Logistics Management,* 24 (10), 20-32.

Bertelsen, S. and Koskela, L. (2002). "Managing the three aspects of production in construction". In: Formoso, C.T. (ed.). *Proceedings 10th Annual IGLC Conference.* UFRGS, Porto Alegre.

Bidault, F., Despres, C. and Butler, C. (1998). "New product development and early supplier involvement (ESI): the drivers of ESI adoption". *International Journal of Technology Management,* 15 (1/2), 49-69.

Binder, M. and Clegg, B. (2007). "Designing and managing collaborative enterprises in the automotive industry". *International Journal of Logistics Research and Applications,* 10 (2), 135-152.

Bititci, U.S., Martinez, V., Albores, P. and Mendibil, K. (2003). "Creating and sustaining competitive advantage in collaborative systems: the what and the how". *Production Planning & Control,* 14 (5), 410-424.

Blismas, N., Sher, W., Thorpe, A. and Baldwin, A. (2004a). "A typology for clients' multi-project environments". *Construction Management and Economics,* 22 (4), 357-371.

Blismas, N.G., Sher, W.D., Thorpe, A. and Baldwin, A.N. (2004b). "Factors influencing project delivery within construction clients' multi-project environments". *Engineering, Construction and Architectural Management,* 11 (2), 113-125.

Blois, K.J. (1972). "Vertical quasi-integration". *Journal of Industrial Economics,* 20, 253-272.

Boardman, J.T. and Clegg, B.T. (2001). "Structured engagement in the extended enterprise". *International Journal of Operations & Production Management,* 21 (5/6), 795-811.

Bossink, B.A.G. and Vrijhoef, R. (2009). "Innovation management in the construction supply chain." In: O'Brien, W.J. et al. (eds.). *Construction Supply Chain Management Handbook*, CRC Press, Boca Raton FL. 14.1-14.21.

Botter, C.H. (1980). *Industrie en organisatie: een overzicht en uitzicht.* Edition 11. Kluwer / NIVE, 267 p.

Bouwend Nederland (2009). "Ketenintegratie haalt prijsangel uit bouwproces". *Podium,* 16 (22 October 2009), 3 p.

Bowersox, D.J. and Closs, D.J. (1996). *Logistical management: the integrated supply chain process.* Edition 3. McGraw-Hill, New York. 730 p.

Bowersox, D.J., Closs, D.J. and Cooper, M.B. (2002). *Supply chain logistics management.* McGraw-Hill, Boston.

Brady, T., Davies, A. and Gann, D. (2005). "Can integrated solutions business models work in construction?". *Building Research & Information,* 33 (6), 571-579.

Bresnen, M. and Marshall, N. (2000a). "Building partnerships: case studies of client–contractor collaboration in the UK construction industry". *Construction Management and Economics,* 18 (7), 819-832.

Bresnen, M. and Marshall, N. (2000b). "Motivation, commitment and the use of incentives in partnerships and alliances". *Construction Management and Economics,* 18, 587-598.

Bresnen, M. and Marshall, N. (2000c). "Partnering in construction: a critical review of issues, problems and dilemmas". *Construction Management and Economics,* 18, 229-237.

Bresnen, M. and Marshall, N. (2002). "The engineering or evolution of cooperation: a tale of two partnering projects". *International Journal of Project Management,* 20, 497-505.

Brewer, J. and Hunter, A. (1989). *Multimethod research: a synthesis of styles*. Sage Publications, Newbury Park. 209 p.

Bridge, A.J. (2004). "The make-or-buy decision in relation to mechanical building services maintenance: a test of an integrated framework of vertical integration". In: *Proceedings COBRA.* 7 September 2004, RICS Foundation, Leeds. 11 p.

Bridge, A.J. (2005). "A make-or-buy decision by mechanical services contractors: a further test of an integrative framework of vertical integration". In: Sidwell, A.C. and Betts, M. (eds.). *Proceedings QUT Research Week.* 4 July 2005, Brisbane. Queensland University of Technology, Brisbane. 12 p.

Briscoe, G. and Dainty, A. (2005). "Construction supply chain integration: an elusive goal?". *Supply Chain Management: An international Journal,* 10 (4), 319-326.

Briscoe, G.H., Dainty, A.R.J. and Millett, S.J. (2001). "Construction supply chain partnerships: skills, knowledge and attitudinal requirements". *European Journal of Purchasing and Supply Management,* 7, 243-255.

Briscoe, G.H., Dainty, A.R.J., Millett, S.J. and Neale, R.H. (2004). "Client-led strategies for construction supply chain improvement". *Construction Management and Economics,* 22 (2), 193-201.

Broersma, H. (1991). *Co-makership: trend voor de jaren negentig*. Rijksuniversiteit Groningen, Groningen.

Building Business (2005a). "Bouwen is systeemintegratie". *Building Business,* 2005 (May), 4 p.

Building Business (2005b). "Ik wil aantonen dat ketenmanagement bijdraagt aan de reductie van faalkosten". *Building Business,* 2005 (August), 5 p.

Building Business (2007a). "Corporaties en ketensamenwerking: op zoek naar de doorbraak". *Building Business,* 2007 (November), 60-63.

Building Business (2007b). "Ketenintegratie in de woningbouw: zo kan het!". *Building Business,* 2007 (February), 50-51.

Building Business (2008). "Bouwbedrijven en ketensamenwerking: op zoek naar de doorbraak (2)". *Building Business,* 2008 (August), 54-59.

Building Business (2009). "Ketenintegratie wordt ter hand genomen: bouwers en corporaties samen in de ring". *Building Business,* 2009 (February), 20-23.

Building Business (2010a). "Ketenintegratie in de nieuwbouw levert winst op". *Building Business,* 2010 (April), 12-14.

Building Business (2010b). "Ketenintegratie ook bij renovatie?: kansen op succes lijken groter dan bij nieuwbouw". *Building Business,* 2010 (April), 16-19.

Burgess, K., Singh, P.J. and Koroglu, R. (2006). "Supply chain management: a structured literature review and implications for future research". *International Journal of Operations & Production Management,* 26 (7), 703-729.

Buzzell, R.D. (1983). "Is vertical integration profitable?". *Harvard Business Review,* September-October, 92-102.

Bygballe, L.E., Jahre, M. and Swärd, A. (2010). "Partnering relationships in construction: a literature review". *Journal of Purchasing & Supply Management,* 16 (2010), 239-253.

Cagliano, R., Caniato, F., Corso, M. and Spina, G. (2005). "Collaborative improvement in the extended manufacturing enterprise: lessons from an action research process". *Production Planning & Control,* 16 (4), 345-355.

Cagliano, R., Caniato, F. and Spina, G. (2006). "The linkage between supply chain integration and manufacturing improvement programmes". *International Journal of Operations & Production Management,* 26 (3), 282-299.

Caldwell, N. and Down, S. (1997). "'Faking it' in supply chain relationships". *Management Research News,* 20 (2), 53-61.

Campbell, J. and Sankaran, J. (2005). "An inductive framework for enhancing supply chain integration". *International Journal of Production Research,* 43 (16), 3321-3351.

Cavinato, J.L. (1992). "A total cost/value model for supply chain competitiveness". *Journal of Business Logistics,* 13 (2), 285-301.

Chandra, C. and Kamur, S. (2000). "An application of a system analysis methodology to manage logistics in a textile supply chain". *Supply Chain Management: An international Journal,* 5 (5), 234-244.

Chandra, C. and Kumar, S. (2000). "Supply chain management in theory and practice: a passing fad or a fundamental change?". *Industrial Management & Data Systems,* 100 (3), 100-113.

Chandra, C. and Kumar, S. (2001). "Taxonomy of inventory policies for supply-chain effectiveness". *International Journal of Retail & Distribution Management,* 29 (4), 164-175.

Chandran, P.M. (2003). *Wall-mart's supply chain management practices.* ICFAI Center for Management Research, Hyderabad India. 13 p.

Chang, C.Y. (2006). "The determinants of the vertical boundaries of the construction firm: comment". *Construction Management and Economics,* 24, 229-232.

Chauhan, S.S. and Proth, J.M. (2005). "Analysis of a supply chain partnership with revenue sharing". *International Journal of Production Economics,* 97 (2005), 44-51.

Chen, T.J. (2003). "Network resources for internationalization: the case of Taiwan's electronics firms". *Journal of Management Studies,* 40 (5), 1107-1130.

Cherns, A.B. and Bryant, D.T. (1984). "Studying the client's role in construction management". *Construction Management and Economics,* 2, 177-184.

Childerhouse, P., Hermiz, R., Mason-Jones, R., Popp, A. and Towill, D.R. (2003a). "Information flow in automotive supply chains: identifying and learning to overcome barriers to change". *Industrial Management & Data Systems,* 103 (7), 491-502.

Childerhouse, P., Lewis, J., Naim, M. and Towill, D.R. (2003b). "Re-engineering a construction supply chain: a material flow control approach". *Supply Chain Management: An international Journal,* 8 (4), 395-406.

Choi, T.Y., Dooley, K.J. and Rungtusanatham, M. (2001). "Supply networks and complex adaptive systems: control versus emergence". *Journal of Operations Management,* 19, 351-366.

Christopher, M. (1992). *Logistics and supply chain management: strategies for reducing cost and improving service.* Pitman / Prentice Hall, London.

Christopher, M. (1997). *Marketing logistics.* Butterworth Heinemann, Oxford.

Christopher, M. and Towill, D. (2000). "Supply chain migration from lean and functional to agile and customised". *Supply Chain Management: An international Journal,* 5 (4), 206-213.

Cobouw (2007). "Steeds meer bedrijven nemen rol van systeemintegrator op zich". *Cobouw,* 2007 (26 April 2007).

Cobouw (2010). "Ketensamenwerking bittere noodzaak". *Cobouw,* 2010 (20 January 2010).

Cooke-Davies, T.J. and Arzymanow, A. (2003). "The maturity of project management in different industries: an investigation into variations between project management models". *International Journal of Project Management,* 21, 471-478.

Cooper, C.L. and Rousseau, D.M. (1999). *Trends in organisational behavior.* John Wiley & Sons, New York.

Cooper, M.C., Douglas, D.M. and Pagh, J.D. (1997). "Supply chain management: more than a new name for logistics". *International Journal of Logistics Management,* 8 (1), 1-14.

Cooper, M.C. and Ellram, L.M. (1993). "Characteristics of supply chain management and the implications for purchasing and logistics strategy". *International Journal of Logistics Management,* 4 (2), 13-24.

Cooper, M.C. and Gardner, J.T. (1993). "Good business relationships: more than just partnerships or strategic alliances?". *International Journal of Physical Distribution and Logistics Management,* 23 (6), 14-26.

Cooper, R., Bruce, M., Wootton, A., Hands, D. and Daly, L. (2003). "Managing design in the extended enterprise". *Building Research & Information,* 31 (5), 367–378 p.

Cooper, R., Hinks, J., Allen, S. and Carmichael, S. (1996). *Adversaries or partners?: a case study of an established long-term relationship between a client and a major contractor.* University of Salford, Salford.

Cousins, P.D., Lawson, B. and Squire, B. (2006). "Supply chain management: theory and practice - the emergence of an academic discipline?". *International Journal of Operations & Production Management,* 26 (7), 697-702.

Cousins, P.D. and Menguc, B. (2006). "The implications of socialization and integration in supply chain management". *Journal of Operations Management,* 24 (5), 604-620.

Cox, A. (1996). "Relational competence and strategic procurement management". *European Journal of Purchasing and Supply Management,* 2 (1), 57-70.

Cox, A. (2001). "Understanding buyer and supplier power: a framework for procurement and supply competence". *Journal of Supply Chain Management,* 37, 8-15.

Cox, A. (2004). "The art of the possible: relationship management in power regimes and supply chains". *Supply Chain Management: An international Journal,* 9 (5), 346-356.

Cox, A. and Ireland, P. (2001). "Managing construction supply chains: the common sense approach for project-based procurement". In: *Proceedings 10th International Annual IPSERA Conference.* 201-213.

Cox, A. and Townsend, M. (1998). *Strategic procurement in construction: towards better practice in the management of construction supply chains.* Telford, London. 346 p.

Cox, A., Watson, G., Lonsdale, C. and Sanderson, J. (2004). "Managing appropriately in power regimes: relationship and performance management in 12 supply chain cases". *Supply Chain Management: An international Journal,* 9 (5), 357-371.

Crane, D. (1997). "Globalization, organizational size, and innovation in the French luxury fashion industry: production of culture theory revisited". *Poetics,* 24, 393-414.

Crespin-Mazet, F. and Ghauri, P. (2007). "Co-development as a marketing strategy in the construction industry". *Industrial Marketing Management,* 36 (2007), 158-172.

Creswell, J.W. (2003). *Research design: qualitative, quantitative, and mixed methods approaches.* Edition Second edition. Sage Publications, Thousand Oaks. xxvi-246 p.

Crook, T.R. and Combs, J.G. (2007). "Sources and consequences of bargaining power in supply chains". *Journal of Operations Management,* 25 (2), 546-555.

Da Villa, F. and Panizzolo, R. (1996). "Buyer-subcontractor relationships in the Italian clothing industry: an interpretive framework". *International Journal of Operations & Production Management,* 16 (7), 38-61.

Dahlman, C.J. (1979). "The problem of externality". *Journal of Law and Economics,* 22, 141-162.

Dainty, A.R.J., Briscoe, G.H. and Millett, S.J. (2001a). "New perspectives on construction supply chain integration". *Supply Chain Management: An international Journal,* 6 (4), 163-173.

Dainty, A.R.J., Briscoe, G.H. and Millett, S.J. (2001b). "Subcontractor perspectives on supply chain alliances". *Construction Management and Economics,* 19 (8), 841-848.

Davis, T. (1993). "Effective supply chain management". *Sloan Management Review,* 34 (4), 35-46.

De Jong, W.S. (2010). *New SCOP method in ASML supply chain: application of enhanced synchronized base stock in planning of supply chain planning environment.* TUE Department Technology Management, Eindhoven. 59 p.

De Ridder, H.A.J. (1994). *Design and construct of complex civil engineering systems: a new approach to organization and contracts.* Delft University Press, Delft. 391 p.

Dekker, K. (1998). "Open building systems: a case study". *Building Research & Information,* 26 (5), 311-318.

Demeter, K. and Gelei, A. (2003). "Supply chain management framework: dimensions and development stages". *International Federation of Purchasing and Supply Management (IFPSM),* Retrieved 6 October 2010 from www.iimm.org. 10 p.

Deming, W.E. (1986). *Out of the crisis.* MIT Press, Cambridge MA.

Díez-Vial, I. (2007). "Explaining vertical integration strategies: market power, transactional attributes and capabilities". *Journal of Management Studies,* 44 (6), 1017-1040.

Dimitriadis, N.I. and Koh, S.C.L. (2005). "Information flow and supply chain management in local production networks: the role of people and information systems". *Production Planning & Control,* 16 (6), 545-554.

Dooley, K.J. (1997). "A complex adaptive systems model of organization change". *Nonlinear Dynamics, Psychology and Life Sciences,* 1 (1), 69-97.

Doorzicht (2000). "Ketenlogistiek binnen de bouw en toeleverende industrie". *Doorzicht,* 2 (Fall 2000), 16-19.

Doran, D. (2003). "Supply chain implications of modularization". *International Journal of Operations & Production Management,* 23 (3), 316-326.

Doran, D., Hill, A., Hwang, K.S. and Jacob, G. (2007). "Supply chain modularisation: Cases from the French automobile industry". *International Journal of Production Economics,* 106 (1), 2-11.Dowlatshahi, S. (1998). "Implementing early supplier involvement: a conceptual framework". *International Journal of Operations & Production Management,* 18 (2), 143-167.

Dowlatshahi, S. (1999). "Early supplier involvement: theory versus practice". *International Journal of Production Research,* 37 (18), 4119-4139.

Drucker, P.F. (1963). *The practice of management.* Heinemann, London.

DTI (2003). *Forging new supply chains in construction.* Department of Trade and Industry, London. 43 p.

Dubois, A. and Gadde, L.E. (2002). "The construction industry as a loosely coupled system: implications for productivity and innovation". *Construction Management and Economics,* 20 (7), 621-631.

Dyer, J.H. and Nobeoka, K. (2000). "Creating and managing a high-performance knowledge-sharing network: the Toyota case". *Strategic Management Journal,* 21, 345-367.

Ebers, M. and Jarillo, J.C. (1998). "The construction, forms, and consequences of industry networks". *International Studies of Management and Organisation,* 27 (4), 3-21.

Eccles, R.G. (1981). "The quasi-firm in the construction industry". *Journal of Economic Behavior and Organization,* 2 (4), 335-357.

Editor (2005). "How Zara fashions its supply chain". *Strategic Direction,* 21 (10), 28-31.

Egan, J. (1998). *Rethinking construction: the report of the construction task force.* DETR, London.

Egan, J. (2002a). *Accelerating change: a report by Strategic Forum for Construction.* Rethinking Construction, London. 42 p.

Egan, J. (2002b). *Rethinking construction: 2002.* Rethinking Construction, London. 21 p.

Egbu, C.O. and Tong, M.K.L. (2003). "Knowledge and communication flow in construction networks and supply chains: challenges and opportunities". In: *Proceedings CIB Student Chapters Symposium.* 26 September 2003, Hong Kong Polytechnic University, Hong Kong.

EIB (2005). *Procesintegratie en innovatief ondernemerschap in de bouwnijverheid.* Economisch Instituut voor de Bouwnijverheid, Amsterdam. 80 p.

EIB (2006a). *Procesintegratie en innovatief ondernemerschap in de aan de bouw toeleverende industrie en de bouwmaterialenhandel.* Economisch Instituut voor de Bouwnijverheid, Amsterdam. 60 p.

EIB (2006b). *Procesintegratie en innovatief ondernemerschap in ontwerpbureaus.* Economisch Instituut voor de Bouwnijverheid, Amsterdam. 92 p.

EIB (2009). *Verwachtingen bouwproductie en werkgelegenheid 2009.* Economisch Instituut voor de Bouwnijverheid, Amsterdam. 116 p.

Eisenhardt, K.M. (1989). "Building theories from case study research". *Academy of Management Review,* 14 (4), 532-550.

Ekstrom, M.A., Bjornsson, H.C., Kunz, J.C., Levitt, R.E. and Nass, C.I. (2003). *The impact of rating systems on subcontracting decisions. a transaction cost analysis.* CIFE Technical Report # 135. Stanford University.

Ellram, L.M. (1990). “Supply chain management: the industrial organization perspective”. *International Journal of Physical Distribution and Logistics Management,* 21 (1), 13-22.

Ellram, L.M. and Cooper, M.C. (1990). “Supply chain management partnerships, and the shipper-third party relationship”. *International Journal of Logistics Management,* 1 (2), 1-10.

Emmett, S. and Crocker, B. (2006). *The relationship-driven supply chain: creating a culture of collaboration throughout the chain.* Gower Publishing, 187 p.

Emory, C.W. and Cooper, D.R. (1991). *Business research methods.* Edition 4. Irwin, Boston. 760 p.

Eng, T.Y. (2003). “The role of e-marketplaces in supply chain management”. *Industrial Marketing Management,* 5567, 1-9.

Fawcett, S.E. and Magnan, G.M. (2002). “The rhetoric and reality of supply chain integration”. *International Journal of Physical Distribution and Logistics Management,* 32 (5), 339-361.

Fellows, R. (2009). “Culture in supply chains.” In: Pryke, S. (ed.). *Construction supply chain management: concepts and case studies,* Wiley-Blackwell, Chichester. 42-72.

Fernie, S., Root, D. and Thorpe, A. (2000). “Supply chain management: theoretical construct for construction”. In: Santiago, Chile. 541-556.

Fernie, S. and Thorpe, A. (2007). “Exploring change in construction: supply chain management”. *Engineering, Construction and Architectural Management,* 14 (4), 319-333.

Flynn, B.B. and Flynn, E.J. (2005). “Synergies between supply chain management and quality management: emerging implications”. *International Journal of Production Research,* 43 (16), 3421-3436.

Flyvbjerg, B. (2006). “Five misunderstandings about case-study research”. *Qualitative Inquiry,* 12 (2), 219-245.

Ford, H. (1922). *My life and work.* Heinemann, London.

Ford, H. (1926). *Today and tomorrow.* Doubleday, New York.

Forza, C. and Vinelli, A. (2000). “Time compression in production and distribution within the textile-apparel supply chain”. *Integrated Manufacturing Systems,* 11 (2), 138-146.

Fransoo, J.C. and Rutten, W.G.M.M. (1994). "A typology of production control situations in process industries". *International Journal of Operations & Production Management,* 14 (12), 47-57.

Frederiksson, P. (2006). "Mechanisms and rationales for the coordination of a modular assembly system: the case of Volvo Cars". *International Journal of Operations & Production Management,* 26 (4), 350-370.

Frohlich, M.T. and Westbrook, R. (2001). "Arcs of integration: an international study of supply chain strategies". *Journal of Operations Management,* 19, 185-200.

Fujimoto, H. (2003). "Collaborative networking in a multi-stage industrial channel". *International Journal of Physical Distribution & Logistics Management,* 33 (3), 229-235.

Fynes, B. and Ainamo, A. (1998). "Organisational learning and lean supply relationships: the case of Apple Ireland". *Supply Chain Management,* 3 (2), 96-107.

Fynes, B., De Burca, S. and Voss, C. (2005a). "Supply chain relationship quality, the competitive environment and performance". *International Journal of Production Research,* 43 (16), 3303-3320.

Fynes, B., Voss, C. and De Burca, S. (2005b). "The impact of supply chain relationship quality on quality performance". *International Journal of Production Economics,* 96, 339-354.

Gadde, L.E. and Jellbo, O. (2002). "System sourcing opportunities and problems". *European Journal of Purchasing & Supply Management,* 8, 43-51.

Galbraith, J.R. (1995). *Designing organizations.* Jossey-Bass, San Francisco.

Gann, D.M. (1996). "Construction as a manufacturing process?: similarities and differences between industrialized housing and car production in Japan". *Construction Management and Economics,* 14, 437-450.

Garg, A. (1999). "An application of designing products and processes for supply chain management". *IIE Transactions,* 31, 417-429.

Gattorna, J.L. (1998). *Strategic supply chain alignment: best practices in supply chain management.* Edition 6. Gower Publishing, Aldershot. 671 p.

Gereffi, G. (1999). "International trade and industrial upgrading in the apparel commodity chain". *Journal of International Economics,* 48, 37-70.

Giannoccaro, I. and Pontrandolfo, P. (2004). "Supply chain coordination by revenue sharing contracts". *International Journal of Production Economics,* 89 (2004), 131-139.

Gibbons, M., Limoges, C., Nowotny, H., Schwartzman, S., Scott, P. and Trow, M. (1994). *The new production of knowledge: the dynamics of science and research in contemporary societies*. Sage Publications, London. 179 p.

Giunipero, L.C. and Brand, R.R. (1996). "Purchasing's role in supply chain management". *International Journal of Logistics Management,* 7 (1), 29-37.

Grandori, A. (1999a). *Interfirm networks: organization and industrial competitiveness*. Routledge, London.

Grandori, A. (1999b). "Interfirm networks: organizational mechanisms and economic outcomes." In: Grandori, A. (ed.). *Interfirm networks: organization and industrial competitiveness*, Routledge, London. 1-14.

Green, K.W. and Inman, R.A. (2005). "Using a just-in-time selling strategy to strengthen supply chain linkages". *International Journal of Production Research,* 43 (16), 3437-3453.

Green, S. and Lenard, D. (1999). "Organising the project procurement process." In: Rowlinson, S. and McDermott, P. (eds.). *Procurement Systems: A Guide to Best Practice in Construction*, Spon, London. 57-82.

Green, S.D., Fernie, S. and Weller, S. (2005). "Making sense of supply chain management: a comparative study of aerospace and construction". *Construction Management and Economics,* 23, 579-593.

Green, S.D., Newcombe, R., Fernie, S. and Weller, S. (2004). *Learning across business sectors: knowledge sharing between aerospace and construction*. University of Reading, Reading. 84 p.

Green, S.D., Newcombe, R., Williams, M., Fernie, S. and Weller, S. (2002). "Supply chain management: a contextual analysis of aerospace and construction". In: Lewis, T.M. (ed.). *Proceedings CIB W092 Symposium: Procurement Systems and Technology Transfer*. University of the West Indies, Trinidad & Tobago. 245-261.

Groák, S. (1994). "Is construction an industry?: note towards a greater analytical emphasis on external linkages". *Construction Management and Economics,* 12, 287-293.

Gummeson, E. (1991). *Qualitative methods in management research*. Sage Publications, Newbury Park. 212 p.

Gundlach, G.T., Bolumole, Y.A., Eltantawy, R.A. and Frankel, R. (2006). "The changing landscape of supply chain management, marketing channels of distribution, logistics and purchasing". *Journal of Business & Industrial Marketing,* 21 (7), 428-438.

Håkansson, H. and Johansson, J. (1994). "A model of industrial networks." In: Axelsson, B. and Easton, G. (eds.). *Industrial networks: a new view of reality*, Routledge, London. 28-34.

Håkansson, H. and Sharma, H.D. (1996). "Strategic alliances in a network perspective." In: Iacobucci, D. (ed.). *Networks in marketing*, Sage, New Dehli. 108-124.

Håkansson, H. and Snehota, I. (1995). *Developing relationships in networks*. Routledge, London.

Haksever, A.M. and Pickering, G. (1996). "Benefits of long-term relationships: UK contractors' experience." In: Langford, D.A. and Retik, A. (eds.). *The organization and management of construction: shaping theory and practice*, Volume 1. Spon, London. 239-246.

Hall, M., Holt, R. and Graves, A. (2000). "Private finance, public roads: configuring the supply chain in PFI highway construction". *European Journal of Purchasing and Supply Management,* 6, 227-235.

Halman, J.I.M., Hofer, A.P. and Van Vuuren, W. (2003). "Platform-driven development of product families: linking theory with practice". *Journal of Product Innovation Management,* 2003 (20), 149-162.

Hammel, T., Kuettner, D. and Phelps, T. (2002). "The re-engineering of Hewlett-Packard's CD-RW supply chain". *Supply Chain Management: An international Journal,* 7 (3), 113-118.

Hammuda, I. and Dulaimi, M.F. (1997). "The theory and application of empowerment in construction: a comparative study of the different approaches to empowerment in construction, service and manufacturing industries". *International Journal of Project Management,* 15 (5), 289-296.

Handfield, R.B. and Nichols, E.L. (2002). *Supply chain redesign: transforming supply chains into integrated value systems*. FT Press, Upper Saddle River NJ. 371 p.

Hannon, D. (2008). "Shorter is better for Toyota's supply chain: automaker aims to localize production and supply base". *Purchasing,* 2008 (14 August 2008).

Harary, F. and Norman, R.Z. (1953). *Graph theory as a mathematical model in social science.* Institute for Social Research, University of Michigan, Ann Arbor. 45 p.

Harland, C.M. (1996). "Supply chain management: relationships, chains and networks". *British Journal of Management,* 7 (Special Issue), 63-80.

Harland, C.M., Lamming, R.C. and Cousins, P.D. (1999). "Developing the concept of supply strategy". *International Journal of Operations & Production Management,* 19 (7), 650-673.

Harrigan, K.R. (1983). *Strategies for vertical integration.* Lexington Books, Lexington.

Harry, M.J. and Schroeder, R. (1999). *Six sigma: the breakthrough management strategy revolutionizing the world's top corporations.* Doubleday Business, New York. 320 p.

Hewitt, F. (1992). "Supply chain integration". In: *Proceedings Annual Conference Council of Logistics Management.* Council of Logistics Management, Oak Brook. 334-341.

Hill, C.A. and Scudder, G.D. (2002). "The use of electronic data interchange for supply chain coordination in the food industry". *Journal of Operations Management,* 20, 375-387.

Hines, P., Rich, N. and Esain, A. (1999). "Value stream mapping: a distribution industry application". *Benchmarking: An International Journal,* 6 (1), 60-77.

Hobbs, J.E. (1996). "A transaction cost approach to supply chain management". *Supply Chain Management: An international Journal,* 1 (2), 15-27.

Hobday, M., Davies, A. and Prencipe, A. (2005). "Systems integration: a core capability of the modern corporation". *Industrial and Corporate Change,* 14 (6), 1109-1143.

Hoekstra, S. and Romme, J. (1992). *Integral logistics structure: developing customer oriented goods flow.* McGraw-Hill, London.

Hoffman, K. (1985). "Clothing, chips and competitive advantage: the impact of microelectronics on trade and production in the garment industry". *World Development,* 13 (3), 371-392.

Holland, C.P. (1995). "Cooperative supply chain management: the impact of inter-organizational information systems". *Journal of Strategic Information Systems,* 4 (2), 117-133.

Holmström, J. (1997). "Product range management: a case study of supply chain operations in the European grocery industry". *Supply Chain Management,* 2 (3), 107-115.

Holti, R., Nicolini, D. and Smalley, M. (2000). *Building down barriers: the handbook of supply chain management.* CIRIA, London. 95 p.

Holweg, M., Disney, S., Holmström, J. and Småros, J. (2005). "Supply chain collaboration: making sense of the strategy continuum". *European Management Journal,* 23 (2), 170-181.

Houlihan, J. (1988). "Exploiting the industrial supply chain." In: Mortimer, J. (ed.). *Logistics in Manufacturing*, IFS Publications, London.

Huang, G.Q., Zhang, X.Y. and Liang, X. (2005). "Towards integrated optimal configuration of platform products, manufacturing processes, and supply chains". *Journal of Operations Management,* 23, 267-290.

Hughes, W., Hillebrandt, P., Greenwood, D. and Kwawu, W. (2006). *Procurement in the construction industry: the impact and cost of alternative market and supply processes.* Taylor & Francis, Abingdon. 208 p.

Imai, M. (1986). *Kaizen: the key to Japan's competitive success.* McGraw-Hill/Irwin, New York.

Ireland, P. (2004). "Managing appropriately in construction power regimes: understanding the impact of regularity in the project environment". *Supply Chain Management: An international Journal,* 9 (5), 372-382.

Ireland, R.D. and Webb, J.W. (2007). "A multi-theoretic perspective on trust and power in strategic supply chains". *Journal of Operations Management,* 25 (2), 482-497.

Iwashita, S. (2001). "Custom made housing in Japan and the growth of the super subcontractor". *Construction Management and Economics,* 19, 295-300.

Iyer, A., Seshadri, S. and Vasher, R. (2009). *Toyota supply chain management: a strategic approach to Toyota's renowned system.* McGraw-Hill, 240 p.

Jiang, B. and Hansen, J.D. (2003). "Matsushita realigns its supply chain in China". *Supply Chain Management: An international Journal,* 8 (3), 185-189.

Johannson, L. (1994). "How can a TQEM approach add value to your supply chain?". *Total quality environmental management,* 3 (4), 521-530.

Johnston, J.E. (1981). *Site control of materials.* Butterworths, London.

Jones, T.C. and Riley, D.W. (1985). "Using inventory for competitive advantage through supply chain management". *International journal of physical distribution and materials management,* 15 (5), 16-26.

Joskow, P.L. (1988). "Asset specificity and the structure of vertical relationships: empirical evidence". *Journal of Law, Economics, and Organization,* 4 (1), 95-117.

Karlsson, C. (2003). "The development of industrial networks: challenges to operations management in an extraprise". *International Journal of Operations & Production Management,* 23 (1), 44-61.

Katzenbach, J.R. and Smith, D.K. (1993). *The wisdom of teams: creating the high-performance organisation.* Harvard Business School Press, Boston.

Kazi, A.S., Hannus, M., Boudjabeur, S. and Malone, A. (2007). *Open building manufacturing: core concepts and industrial requirements.* Manubuild / VTT, Helsinki. 320 p.

Kelly, J. and Male, S. (2001). "A value management approach to aligning the team to the client's value system". In: *Proceedings COBRA.* RICS Foundation.

Kenley, R., London, K. and Watson, J. (2000). "Strategic procurement in the construction industry mechanisms for public sector clients to improve performance in the Australian public sector". *Journal of Construction Procurement,* 6 (1), 4-19.

Khalfan, M.M.A., McDermott, P., Vrijhoef, R. and Asad, S. (2005). "The effect of procurement on the integration of the supply chain within the construction industry". In: Kähkönen, K. and Sexton, M. (eds.). *Proceedings CIB Combining Forces.* Volume 1: Understanding the Construction Business and Companies in the New Millennium. 13 June 2005, Helsinki. VTT, Espoo. 14-25.

Kim, B., Park, K. and Kim, T. (1999). "The perception gap among buyer and suppliers in the semiconductor industry". *Supply Chain Management: An international Journal,* 4 (5), 231-241.

Kim, S.W. (2006). "The effect of supply chain integration on the alignment between corporate competitive capability and supply chain operational capability". *International Journal of Operations & Production Management,* 26 (10), 1084-1107.

King, R.P. and Phumpiu, P.F. (1996). "Reengineering the food supply chain: the ECR initiative in the grocery industry". *American Journal of Agricultural Economics,* 78, 1181-1186.

Kirche, E.T., Kadipasaoglu, S.N. and Khumawala, B.M. (2005). "Maximizing supply chain profits with effective order management: integration of activity-based costing and theory of constraints with mixed-integer modelling". *International Journal of Production Research,* 43 (7), 1297–1311 p.

Klein, P.G. (2005). "The make-or-buy decision: lessons from empirical studies." In: Ménard, C. and Shirley, M.M. (eds.). *Handbook of New Institutional Economics*, Springer, Dordrecht. 435-464.

Kornelius, L. and Wamelink, J.W.F. (1998). "The virtual corporation: learning from construction". *Supply Chain Management: An international Journal,* 3 (4), 193-202.

Koskela, L. (1993). "Lean production in construction". In: *Proceedings 1st Annual IGLC Conference.* VTT, Espoo.

Koskela, L. (2000). *An exploration towards a production theory and its application to construction*. PhD Thesis. VTT Publications 408. VTT Technical Research Centre of Finland, Espoo. 296 p.

Koskela, L., Howell, G.A., Ballard, G. and Tommelein, I. (2002). "The foundations of lean construction." In: Best, R. and De Valence, G. (eds.). *Design and Construction: Building in Value*, Butterworth-Heinemann, Oxford. 211-226.

Koufteros, X.A., Vonderembse, M.A. and Doll, W.J. (2002). "Integrated product development practices and competitive capabilities: the effects of uncertainty, equivocality, and platform strategy". *Journal of Operations Management,* 20, 331-355.

Krikke, H., Bloemhof-Ruwaard, J. and Van Wassenhove, L.N. (2003). "Concurrent product and closed-loop supply chain design with an application to refrigerators". *International Journal of Production Research,* 41 (16), 3689-3719.

Kwon, I.G. and Suh, T. (2005). "Trust, commitment and relationships in supply chain management: a path analysis". *Supply Chain Management: An international Journal,* 10 (1), 26-33.

Lai, K.H. and Cheng, T.C.E. (2003). "Initiatives and outcomes of quality management implementation across industries". *Omega,* 31, 141-154.

Lai, K.H., Cheng, T.C.E. and Yeung, A.C.L. (2005). "Relationship stability and supplier commitment to quality". *International Journal of Production Economics,* 96, 397-410.

Lakshman, C. and Parente, R.C. (2008). "Supplier-focused knowledge management in the automobile industry and its implications for product performance". *Journal of Management Studies,* 45 (2), 317-342.

Lambert, D.M. (2008). *Supply chain management: processes, partnerships, performances.* Edition 3. Supply Chain Management Institute, Sarasota. 431 p.

Lambert, D.M., Emmelhainz, M.A. and Gardner, J.T. (1996). "Developing and implementing supply chain partnerships". *International Journal of Logistics Management,* 7 (2), 1-17.

Lambert, D.M., Knemeyer, A.M. and Gardner, J.T. (2004). "Supply chain partnerships: model validation and implementation". *Journal of Business Logistics,* 25 (2), 21-42.

Lambert, D.M., Pugh, M. and Cooper, J. (1998). "Supply chain management". *International Journal of Logistics Management,* 9 (2), 1-19.

Lamming, R. (1993). *Beyond partnership: strategies for innovation and lean supply.* Prentice Hall, New York. 299 p.

Lamming, R. (1996). "Squaring lean supply with supply chain management". *International Journal of Operations & Production Management,* 16 (2), 183-196.

Lamming, R., Zhang, J., Caldwell, N. and Phillips, W. (2005). "Implementing value-transparency in supply relations". *Management Decision,* 43 (6), 810-820.

Lander, E. and Liker, J.K. (2007). "The Toyota Production System and art: making highly customized and creative products the Toyota way". *International Journal of Production Research,* 45 (16), 3681-3698.

Langford, D.A., Martinez, V. and Bititci, U. (2003). "Best value in construction: towards an interpretation of value from client and constructor perspectives". *Journal of Construction Procurement,* 9 (1).

Langley, A. (1999). "Strategies for theorizing from process data". *Academy of Management Review,* 24 (4), 691-710.

Lansley, P. (1994). "Analysing construction organizations". *Construction Management and Economics,* 12 (4), 337-348.

Lau, A.K.W. and Yam, R.C.M. (2005). "A case study of product modularization on supply chain design and coordination in Hong Kong and China". *Journal of Manufacturing Technology Management,* 16 (4), 432-446.

Lee, H.L. and Billington, C. (1992). "Managing supply chain inventory: pitfalls and opportunities". *Sloan Management Review,* 33 (3), 65-73.

Lee, H.L. and Whang, S. (1999). "Supply chain integration in the age of e-business". *Supply Chain Management Review,* 1999 (Fall), 16-19.

Lee, Y.H. and Kim, T. (2002). "Manufacturing cycle time reduction using balance control in the semiconductor fabrication line". *Production Planning & Control,* 13 (6), 529-540.

Leedy, P.D. and Ormrod, J.E. (2001). *Practical research: planning and design.* Prentice-Hall, Upper Saddle River.

Lesser, E.L. and Storck, J. (2001). "Communities of practice and organizational performance". *IBM Systems Journal,* 40 (4), 831-841.

Levy, P., Bessant, J., Sang, B. and Lamming, R. (1995). "Developing integration through total quality supply chain management". *Integrated Manufacturing Systems,* 6 (3), 4-12.

Li, S., Rao, S.R., Ragu-Nathan, T.S. and Ragu-Nathan, B. (2005). "Development and validation of a measurement instrument for studying supply chain management practices". *Journal of Operations Management,* 23, 618-641.

Lin, F.R. and Shaw, M.J. (1998). "Reengineering the order fulfilment process in supply chain networks". *International Journal of Flexible Manufacturing Systems,* 10, 197-299.

Loader, R. (1997). "Assessing transaction costs to describe supply chain relationships in agri-food systems". *Supply Chain Management: An international Journal,* 1 (2), 23-35.

Lockamy, A. and McCormack, K. (2004). "The development of a SCM process maturity model using the concepts of business process orientation". *Supply Chain Management: An international Journal,* 9 (4), 272-278.

Lockamy, A. and Smith, W.I. (2000). "Target costing for supply chain management: criteria and selection". *Industrial Management & Data Systems,* 100 (5), 210-218.

Loebbecke, C., Kronena, J.H. and Jelassib, T. (1996). "The role of information technology in retailing: the case of supporting fashion purchasing at a European department store chain". *Journal of Strategic Information Systems,* 5, 67-78.

London, K. (2001). "Evolution of a supply chain firm constellation for innovative construction products: an instrumental case study". In: *Proceedings CIB Triennial World Congress.* April 2001, Wellington.

London, K. and Kenley, R. (1999). "Client's role in construction supply chains: a theoretical discussion". In: *Proceedings CIB Triennial World Symposium W92.* Cape Town. 100-111.

London, K. and Kenley, R. (2001). "An industrial organization economic supply chain approach for the construction industry: a review". *Construction Management and Economics,* 19 (8), 777-788.

London, K., Kenley, R. and Agapiou, A. (1998). "The impact of construction industry structure on supply chain network modelling". In: *Proceedings Annual Conference Logistics Research Network.* Cranfield School of Management, Cranfield. 10 p.

Love, P.E.D., Edwards, D.J. and Sohal, A. (2004a). "Total quality management in Australian contracting organisations: pre-conditions for successful implementation". *Engineering, Construction and Architectural Management,* 11 (3), 189-198.

Love, P.E.D., Holt, G.D. and Li, H. (2002a). "Triangulation in construction management research". *Engineering, Construction and Architectural Management,* 9 (4), 294-303.

Love, P.E.D., Irani, Z., Cheng, E. and Li, H. (2002b). "A model for supporting inter-organizational relations in the supply chain". *Engineering, Construction and Architectural Management,* 9 (1), 2-15.

Love, P.E.D., Irani, Z. and Edwards, D.J. (2004b). "A seamless supply chain management model for construction". *Supply Chain Management: An international Journal,* 9 (1), 43-56.

Love, P.E.D., Tse, R.Y.C., Holt, G.D. and Proverbs, D.G. (2002c). "Transaction costs, learning and alliances in construction". *Journal of Construction Research,* 3 (2), 193-207.

Luhtala, M., Kilpinen, E. and Anttila, P. (1994). *LOGI: managing make-to-order supply chains.* Helsinki University of Technology, Espoo.

Lummus, R.R., Krumwiede, D.W. and Vokurka, R.J. (2001). "The relationship of logistics to supply chain management: developing a common industry definition". *Industrial Management & Data Systems,* 101 (8), 426-431.

MacDuffie, J.P. and Helper, S. (1997a). "Creating lean production suppliers: diffusing lean through the supply chain". *California Management Review,* 39 (4), 118-151.

MacDuffie, J.P. and Helper, S. (1997b). "Creating lean production suppliers: diffusing lean through the supply chain". *California Management Review,* 39 (4), 118-151.

Magretta, J. (1998a). "The power of virtual integration: an interview with Dell Computer's Michael Dell". *Harvard Business Review,* 1998 (March-April), 73-84.

Magretta, J. (1998b). "The power of virtual integration: an interview with Dell Computer's Michael Dell". *Harvard Business Review,* 1998 (March-April), 73-84.

Mahoney, J.T. (1992). "The choice of organizational form: vertical financial ownership versus other methods of vertical integration". *Strategic Management Journal,* 13, 559-584.

Maloni, M.J. and Benton, W.C. (1997). "Supply chain partnerships: opportunities for operations research". *European Journal of Operational Research,* 101 (1997), 419-429.

Mariotti, S. and Cainarca, G.C. (1986). "The evolution of transaction governance in the textile-clothing industry". *Journal of Economic Behavior and Organization,* 7, 351-374.

Martinsuo, M. and Lehtonen, P. (2007). "Role of single-project management in achieving portfolio management efficiency". *International Journal of Project Management,* 25 (1), 56-65.

Mason, S.J., Cole, M.H., Ulrey, B.T. and Yan, L. (2002). "Improving electronics manufacturing supply chain agility through outsourcing". *International Journal of Physical Distribution & Logistics Management,* 32 (7), 610-620.

Maxwell, J. (2007). *Types of generalization in qualitative research.* Retrieved 23 September 2010 from www.tcrecord.org.

McAfee, R.B., Glassman, M. and Honeycutt, E.D. (2002). "The effects of culture and human resource management policies on supply chain management strategies". *Journal of Business Logistics,* 23 (1), 1-18.

McMichael, H., Mackay, D. and Altmann, G. (2000). "Quick response in the Australian TCF industry: a case study of supplier response". *International Journal of Physical Distribution & Logistics Management,* 30 (7/8), 611-626.

McNamee, P. and McHugh, M. (1989). "Mapping competitive groups in the clothing industry". *Long Range Planning,* 22 (5), 89-97.

Meijboom, B. (1999). "Production-to-order and international operations: a case study in the clothing industry". *International Journal of Operations & Production Management,* 19 (5/6), 602-619.

Melles, B. and Wamelink, J.W.F. (1993). *Production control in construction: different approaches to control, use of information and automated data processing.* Delft University of Technology, Delft. 320 p.

Merli, G. (1991). *Co-makership: the new supply strategy for manufacturers.* Productivity Press, Cambridge MA.

Meyer, M.H. and Lehnerd, A.P. (1997). *The power of product platforms.* Free Press, New York.

Michaels, L.M.J. (1999). "The making of a lean aerospace supply chain". *Supply Chain Management,* 4 (3), 135-144.

Miles, M.B. and Huberman, A.M. (1994). *Qualitative data analysis: an expanded sourcebook.* Edition 2. Sage, Thousand Oaks. 338 p.

Milgrom, P.R. and Roberts, J. (1990). "The economics of modern manufacturing: technology, strategy, and organization". *American Economic Review,* 80, 511-528.

Milgrom, P.R. and Roberts, J. (1992). *Economics, organizations, and management.* Prentice-Hall, Englewood Cliffs.

Miller, J.B. (1999). "Applying multiple project procurement methods to a portfolio of infrastructure projects." In: Rowlinson, S. and McDermott, P. (eds.). *Procurement Systems: A Guide to Best Practice in Construction*, Spon, London.

Millett, S.J., Dainty, A.R.J., Briscoe, G.H. and Neale, R.H. (2000). "Towards an integrated construction supply network: sub-contractors perspectives". In: Akintoye, A. (ed.). *Proceedings 16th Annual ARCOM Conference*. Volume 2. 6 September 2000, Glasgow Caledonian University, Glasgow. 705-714.

Minegishi, S. and Thiel, D. (2000). "System dynamics modeling and simulation of a particular food supply chain". *Simulation Practice and Theory,* 8, 321-339.

Monden, Y. (1994). *Toyota production system: an integrated approach to just-in-time*. Chapman & Hall, London.

Moon, S.A. and Kim, D.J. (2005). "Systems thinking ability for supply chain management". *Supply Chain Management: An international Journal,* 10 (5), 394-401.

Morris, D., Donnelly, T. and Donnelly, T. (2004). "Supplier parks in the automotive industry". *Supply Chain Management: An international Journal,* 9 (2), 129-133.

Mould, G. and King, M. (1995). "Just-in-time implementation in the Scottish electronics industry". *Industrial Management and Data Systems,* 95 (9), 17-22.

Muffatto, M. (1999). "Introducing a platform strategy in product development". *International Journal of Production Economics,* 60/61, 145-153.

Mukhopadhyay, S.K. and Setoputro, R. (2005). "Optimal return policy and modular design for build-to-order products". *Journal of Operations Management,* 23, 496-506.

Murty, K. G. (). Supply Chain Management in the Computer Industry. 13-10-2010.

Naim, M., Naylor, J. and Barlow, J. (1999). "Developing lean and agile supply chains in the UK housebuilding industry". In: Tommelein, I.D. (ed.). *Proceedings 7th Annual IGLC Conference*. University of California, Berkeley. 159-170.

Nam, C.H. and Tatum, C.B. (1988). "Major characteristics of constructed products and resulting limitations of construction technology". *Construction Management and Economics,* 6, 133-148.

Ndekugri, I. and Corbett, P. (2004). "Supply chain integration in construction by prime contracting: some research issues". In: *Proceedings COBRA*. 7 September 2004, RICS Foundation, Leeds. 14 p.

Nicolini, D., Holti, R. and Smalley, M. (2001). "Integrating project activities: the theory and practice of managing the supply chain through clusters". *Construction Management and Economics,* 19 (1), 37-47.

Nobeoka, K. and Cusumano, M.A. (1997). "Multiproject strategy and sales growth: the benefit of rapid design transfer in new product development". *Strategic Management Journal,* 18, 169-186.

Noorderhaven, N.G., Koen, C.I. and Beugelsdijk, S. (2002). *Organizational culture and network embeddedness.* CentER, Faculty of Economics and Business Administration, Tilburg University, Tilburg. 51 p.

Nooteboom, B. (1999). "The triangle: roles of the go-between." In: Gabbay, S.M. and Leenders, R. (eds.). *Corporate social capital and liability*, Kluwer, Deventer. 341-355.

Novack, R. and Simco, S.W. (1991). "The industrial procurement process: a supply chain perspective". *Journal of Business Logistics,* 12 (1), 145-167.

O'Brien, W.J. and Fischer, M.A. (1993). "Construction supply chain management: a research framework". In: *Proceedings 3rd International Conference Information Technology for Civil and Structural Engineers CIVIL-COMP-'93.* Edinburgh. 61-64.

O'Brien, W.J., Fischer, M.A. and Jucker, J.V. (1995). "An economic view of project coordination". *Construction Management and Economics,* 13 (5), 393-400.

O'Brien, W.J., London, K. and Vrijhoef, R. (2002). "Construction supply chain modeling: a research review and interdisciplinary research agenda". In: Formoso, C.T. (ed.). *Proceedings 10th Annual IGLC Conference.* UFRGS, Porto Alegre. 129-148.

Ohno, T. (1998). *The Toyota production system: beyond large scale production.* Productivity Press, Cambridge MA. 143 p.

Olhager, J. (2002). "Supply chain management: a just-in-time perspective". *Production Planning & Control,* 13 (8), 681-687.

Olsson, F. (2000). *Supply chain management in the construction industry: opportunity or utopia?* Licentiate thesis. Lund University, Lund. 153 p.

Ouchi, W.G. (1980). "Markets, bureaucracies, clans". *Administrative science quarterly,* 25, 129-141.

Parfitt, A. (2003). *Forging new supply chains: partnership sourcing in construction.* Retrieved 15 May 2005 from http://www.lancs.ac.uk/users/scistud/supply_res.htm.

Park, Y.B. (2005). "An integrated approach for production and distribution planning in supply chain management". *International Journal of Production Research,* 43 (6), 1205–1224 p.

Parkhe, A. (1993). "Strategic alliance structuring: a game theoretic and transaction cost examination of inter-firm cooperation". *Academy of Management Journal,* 36, 794-829.

Patton, M.Q. (2002). *Qualitative research and evaluation methods.* Edition 3. Sage Publications, Thousand Oaks. 598 p.

Perry, M. and Sohal, A.S. (2000). "Quick response practices and technologies in developing supply chains: a case study". *International Journal of Physical Distribution & Logistics Management,* 30 (7/8), 627-639.

Perry, M. and Sohal, A.S. (2001). "Effective quick response practices in a supply chain partnership: an Australian case study". *International Journal of Operations & Production Management,* 21 (5/6), 840-854.

Persson, G. and Virum, H. (2001). "Logistics service providers and supply chain alliances". In: *Proceedings 10th International Annual IPSERA Conference.* 707-724.

Petersen, K.J., Handfield, R.B. and Ragatz, G.L. (2005). "Supplier integration into new product development: coordinating product, process and supply chain design". *Journal of Operations Management,* 23, 371-388.

Pheng, L.S. and Chuan, C.J. (2001). "Just-in-time management in precast concrete construction". *Integrated Manufacturing Systems,* 12 (6), 416-429.

Phillips, M. (1999). "Agile manufacturing in the aerospace industry: an industrial viewpoint". *International Journal of Agile Management Systems,* 1 (1), 17-22.

Pietrobelli, C. and Barrera, T.O. (2002). "Enterprise clusters and industrial districts in Colombia's fashion sector". *European Planning Studies,* 10 (5), 541-562.

Porter, M.E. (1985). *Competitive advantage: creating and sustaining superior performance.* Free Press, New York.

Powell, W.W. (1987). "Hybrid organizational arrangements". *California Management Review,* 30, 67-87.

Powell, W.W. (1990). "Neither market nor hierarchy: network forms of organization." In: Staw, B.M. and Cummings, L.L. (eds.). *Research in Organizational Behavior,* Volume 12. JAI-Press, Greenwich. 295-336.

Prahalad, C.K. and Hamel, G. (1990). "The core competence of the corporation". *Harvard Business Review,* 68 (3), 79-91.

Proverbs, D.G. and Holt, G.D. (2000). "Reducing construction costs: European best practice supply chain implications". *European Journal of Purchasing and Supply Management,* 6, 149-158.

Pryke, S.D. (2002). "Construction coalitions and the evolving supply chain management paradox: progress through fragmentation". In: *Proceedings COBRA.* 5 September 2002, Nottingham.

Pryke, S.D. (2004). "Twenty first century procurement strategies: analysing networks of inter-firm relationships". *RICS Foundation Research Paper Series,* 4 (27), 37 p.

Pryke, S.D. (2005). "Towards a social network theory of project governance". *Construction Management and Economics,* 23 (9), 927-939.

Rae-Smith, J.B. and Ellinger, A.E. (2002). "Insights from the introduction of an online logistics service system". *Supply Chain Management: An international Journal,* 7 (1), 5-11.

Rai, R. and Allada, V. (2003). "Modular product family design: agent-based pareto-optimization and quality loss function-based post optimal analysis". *International Journal Production Research,* 41 (17), 4075-4098.

Rama, R., Ferguson, D. and Melero, A. (2003). "Subcontracting networks in industrial districts: the electronics industries of Madrid". *Regional Studies,* 37 (1), 71-88.

Raman, A. (1999). "Managing inventory for fashion products." In: Tayur, S., Ganeshan, R. and Magazine, M. (eds.). *Quantitative models for supply chain management*, Kluwer Academic Publishers, Boston. 789-805.

Ren, J.Z., Cohen, M.A., Terwiesch, C. and Ho, T. (2003). "An empirical analysis of forecast sharing in the semiconductor equipment supply chain". *Manufacturing & Service Operations Management,* 5 (1), 70-73.

Rich, N. and Hines, P. (1997). "Supply-chain management and time-based competition: the role of the supplier association". *International Journal of Physical Distribution & Logistics Management,* 27 (3/4), 210-225.

Riley, M.J. and Clare-Brown, D. (2001). "Comparison of cultures in construction and manufacturing industries". *Journal of Management in Engineering,* 17 (3), 149-158.

Rintala, K. (2004). *The economic efficiency of accommodation service PFI projects.* VTT Publications 555. VTT, Espoo. 286 p.

Robertson, P.L. and Langlois, R.N. (1995). "Innovation, networks and vertical integration". *Research Policy,* 24, 543-562.

Robson, C. (1993). *Real world research: a resource for social scientists and practitioner-researchers.* Blackwell Publishing, Oxford. 510 p.

Root, D., Thorpe, A., Thomson, D.S., Austin, S.A. and Hammond, J.W. (2003). "Design chains: introducing supply chain management into the construction design process". *Journal of Construction Procurement,* 9 (2), 29-39.

Ryu, S., Han, J.H. and Frank, J. (2006). "Does culture matter? Collectivism, long-term orientation, and supply chain management in Korea". *International Journal of Internet and Enterprise Management,* 4 (2), 162-179.

Sahay, B.S. (2003). "Supply chain collaboration: the key to value creation". *Work Study,* 52 (2), 76-83.

Sambasivan, M. and Yen, C.N. (2010). "Strategic alliances in a manufacturing supply chain: influence of organizational culture from the manufacturer's perspective". *International Journal of Physical Distribution & Logistics Management,* 40 (6), 456-474.

Santema, S. and Van de Rijt, J. (2005). "The valuation of resources: looking through the eyes of the customer." In: Sanchez, R. and Heene, A. (eds.). *Competence Perspectives on Resources, Stakeholders and Renewal (Advances in Applied Business Strategy, Volume 9),* Emerald, Bingley. 17-28.

Sarja, A. (1998). *Open and industrialised building.* Taylor & Francis, Oxford. 228 p.

SBR (2007). *Strategie in de bouw: ondernemers inspireren tot het maken van keuzes.* Stichting Bouwresearch, Rotterdam. 61 p.

SBR (2009). "Ketenintegratie in de woningbouw". Stichting Bouwresearch, *B+ Magazine,* 2009 (2 October 2009), 24-25.

SBR (2010a). "Samenwerking volgens ketenintegratie". Stichting Bouwresearch, *B+ Magazine,* 2010 (3 January 2010), 32-34.

SBR (2010b). *Verkenning ketensamenwerking AFNL: samenwerking en ketenintegratie hoofdaannemers en gespecialiseerde aannemers.* Stichting Bouwresearch, Rotterdam. 38 p.

Schmenner, R.W. (1993). *Production/operations management: from the inside out.* Macmillian, New York.

Scott, C. and Westbrook, R. (1991). "New strategic tools for supply chain management". *International Journal of Physical Distribution & Logistics Management,* 21 (1), 23-33.

Segerstedt, A. and Olofsson, T. (2010). "Supply chains in the construction industry". *Supply Chain Management: An international Journal,* 15 (5), 347-353.

Sehdev, K., Fan, I.S., Cooper, S. and Williams, G. (1995). "Design for manufacture in the aerospace extended enterprise". *World class design to manufacture,* 2 (2), 28-33.

Sense, A.J. and Clements, M.D.J. (2006). "Ever consider a supply chain as a "community of practice"?: embracing a learning perspective to build supply chain integration". *Development and Learning in Organizations,* 20 (5), 6-8.

Seuring, S.A. (2008). "Assessing the rigor of case study research in supply chain management". *Supply Chain Management: An International Journal,* 13 (2), 128-137.

Shekhar, S. (2006). "Understanding the virtuality of virtual organizations". *Leadership & Organization Development Journal,* 27 (6), 465-483.

Shelanski, H.S. and Klein, P.G. (1995). "Empirical research in transaction cost economics: a review and assessment". *Journal of Law, Economics & Organization,* 11 (2), 335-361.

Shub, A.N. and Stonebraker, P.W. (2009). "The human impact on supply chains: evaluating the importance of "soft" areas on integration and performance". *Supply Chain Management: An international Journal,* 14 (1), 31-40.

Silverman, D. (2000). *Doing qualitative research: a practical handbook.* Sage Publications Ltd, London. 316 p.

Simatupang, T.M., Wright, A.C. and Sridharan, R. (2002). "The knowledge of coordination for supply chain integration". *Business Process Management Journal,* 8 (3), 289-308.

Skipworth, H. and Harrison, A. (2006). "Implications of form postponement to manufacturing a customized product". *International Journal of Production Research,* 44 (8), 1627-1652.

Skjoett-Larsen, T., Thernoe, C. and Andresen, C. (2003). "Supply chain collaboration: theoretical perspectives and empirical evidence". *Distribution & Logistics Management,* 33 (6), 531-549.

Smaling, A. (2003). "Inductive, analogical, and communicative generalization". *International Journal of Qualitative Methods,* 2 (1), Article 5. Retrieved 23 September 2010 from http://www.ualberta.ca/~iiqm/backissues/2_1/html/smaling.html. 12 p.

Snijders, C. and Vos, B. (2007). "Whatever the method, it needs to be well-defined, well-argued, and well-executed". *Journal of Purchasing and Supply Management,* 13 (3), 169-170.

Sohal, A.S., Perry, M. and Pratt, T. (1998). "Developing partnerships and networks: learning from practices in Australia". *Technovation,* 18 (4), 245-251.

Soosay, C.A., Hyland, P.W. and Ferrer, M. (2008). "Supply chain collaboration: capabilities for continuous innovation". *Supply Chain Management: An international Journal,* 13 (2), 160-169.

Sparks, L. and Wagner, B.A. (2003). "Retail exchanges: a research agenda". *Supply Chain Management: An international Journal,* 8 (1), 17-25.

Stake, R.E. (1995). *The art of case study research.* Sage Publications, Thousand Oaks. 175 p.

Stank, T.P., Keller, S.B. and Daugherty, P.J. (2001). "Supply chain collaboration and logistical service performance". *Journal of Business Logistics,* 22 (1), 29-48.

Stevens, G.C. (1990). "Successful supply-chain management". *Management Decisions,* 28 (8), 25-30.

Stevenson, M., Hendry, L.C. and Kingsman, B.G. (2005). "A review of production planning and control: the applicability of key concepts to the make-to-order industry". *International Journal of Production Research,* 43 (5), 869–898 p.

Stock, G.N., Greis, N.P. and Kasarda, J.P. (2000). "Enterprise logistics and supply chain structure: the role of fit". *Journal of Operations Management,* 18 (5), 531-547.

Stonebraker, P.W. and Liao, J. (2006). "Supply chain integration: exploring product and environmental contingencies". *Supply Chain Management: An international Journal,* 11 (1), 34-43.

Strauss, A. and Corbin, J. (1998). *Basics of qualitative research: techniques and procedures for developing grounded theory.* Edition 2. Sage Publications Inc, Thousand Oaks.

Stuckey, J. and White, D. (1993). "When and when not to vertically integrate". *Sloan Management Review,* spring, 71-83.

Sturgeon, T.J. (2002). "Modular production networks: a new American model of industrial organization". *Industrial and Corporate Change,* 11 (3), 451-496.

Sturgeon, T.J. (2003). "What really goes on in silicon valley?: spatial clustering and dispersal in modular production networks". *Journal of Economic Geography,* 3, 199-225.

Su, J.C.P., Chang, Y.L. and Ferguson, M. (2005). "Evaluation of postponement structures to accommodate mass customization". *Journal of Operations Management,* 23, 305-318.

Surana, A., Kumara, S., Greaves, M. and Raghavan, U.N. (2005). "Supply-chain networks: a complex adaptive systems perspective". *International Journal of Production Research,* 43 (20), 4235-4265.

Swanson, R.A. and Holton, E.F. (2005). *Research in organizations: foundations and methods of inquiry.* Berrett-Koehler Publishers Inc., San Francisco. xvii-459 p.

Sydow, J. and Windeler, A. (1998). "Organizing and evaluating interfirm networks: a structurationist perspective on network processes and effectiveness". *Organization Science,* 9 (3), 265-284.

Tatsiopoulos, I.P., Panayiotou, N.A. and Ponis, S.T. (2002). "A modelling and evaluation methodology for e-commerce enabled BPR". *Computers in Industry,* 49, 107-121.

Taylor, J.R. (1993). *Rethinking the theory of organizational communication.* Ablex, Norwood.

Teece, D.J. (1992). "Competition, cooperation and innovation". *Journal of Economic Behavior and Organisation,* 18, 1-25.

Tommelein, I. (1998). "Pull-driven scheduling for pipe-spool installation: simulation of lean construction technique". *Journal of Construction Engineering and Management,* 24 (4), 279-288.

Towill, D.R. (1996a). "Industrial dynamics modeling of supply chains". *International Journal of Physical Distribution & Logistics Management,* 26 (2), 23-42.

Towill, D.R. (1996b). "Industrial dynamics modelling of supply chains". *Logistics Information Management,* 9 (4), 43-56.

Towill, D.R. (1996c). "Time compression and supply chain management a guided tour". *Supply Chain Management,* 1 (1), 15-27.

Towill, D.R. (2007). "Exploiting the DNA of the Toyota Production System". *International Journal of Production Research,* 45 (16), 3619-3637.

Towill, D.R. and Childerhouse, P. (2006). "Enabling the seamless supply chain by exploiting the four smooth material flow controls". *Production Planning & Control,* 17 (8), 756-768.

Trkman, P., Štemberger, M.I., Jaklic, J. and Groznik, A. (2007). "Process approach to supply chain integration". *Supply Chain Management: An international Journal,* 12 (2), 116-128.

Tsang, E.W.K. (2000). "Transaction cost and resource-based explanation of joint-ventures: a comparison and synthesis". *Organization Studies,* 21, 215-242.

Tukel, O.I. and Rom, W.O. (1998). "Analysis of the characteristics of projects in diverse industries". *Journal of Operations Management,* 16, 43-61.

Turin, D.A. (2003). "Building as a process". *Building Research & Information,* 31 (2), 180-187. Reprint of the original article first published in the Proceedings of the Bartlett Society in 1967.

Turner, J.R. (1993). "Integrated supply chain management: what's wrong with this picture?". *Industrial Engineering,* 25 (12), 52-55.

Turner, J.R. (2006). "Towards a theory of project management: the functions of project management". *International Journal of Project Management,* 24, 187-189.

Tyan, J. and Wee, H.M. (2003). "Vendor managed inventory: a survey of the Taiwanese grocery industry". *Journal of Purchasing and Supply Management,* 9 (1), 11-18.

Van Aken, J.E. (2005). "Management research as a design science: articulating the research products of mode 2 knowledge production in management". *British Journal of Management,* 16 (1), 19-36.

Van der Vorst, J.G.A.J. (2000). *Effective food supply chains: generating, modelling and evaluating supply chain scenarios.* Dissertation. Wageningen University, Wageningen.

Van Hoek, R.I. (1997). "Postponed manufacturing: a case study in the food supply chain". *Supply Chain Management,* 2 (2), 63-75.

Van Hoek, R.I. (1999). "Postponement and the reconfiguration challenge for food supply chains". *Supply Chain Management,* 4 (1), 18-34.

Van Randen, A. (1990). "Separation of support and infill: a chance for a quantum leap in productivity". In: *Proceedings Open Industrialisation: A Solution for Building Modernisation.* 21 February 1990, Stuttgart.

Van Reijswoud, V.E. (1996). *The structure of business communication.* Dissertation. Delft University of Technology, Delft.

Visser, E.J. (1999). "A comparison of clustered and dispersed firms in the small-scale clothing industry of Lima". *World Development,* 27 (9), 1553-1570.

Voordijk, J.T. (1999). "Preconditions and dynamics of logistics networks in the Dutch building industry". *Supply Chain Management: An international Journal,* 4 (3), 145-154.

Voordijk, J.T. (2000). "The changing logistical system of the building of materials supply chain". *International Journal of Operations & Production Management,* 20 (7), 823-841.

Voordijk, J.T., De Haan, J. and Joosten, G.J. (2000). "Changing governance of supply chains in the building industry: a multiple case study". *European Journal of Purchasing and Supply Management,* 6, 217-225.

Voordijk, J.T., Maas, N. and Verduijn, T. (1999). "Changing pipeline management of agricultural products: a case study." In: Bartezzaghi, E. et al. (eds.). *Managing operations networks*, SG Editoriali, Venice. 395-402.

Voordijk, J.T. and Vrijhoef, R. (2003). "Improving supply chain management in construction; what can be learned from the aerospace industry?". In: Greenwood, D.J. (ed.). *Proceedings Annual ARCOM Conference.* Volume 2. 3 September 2003, University of Brighton, Brighton. 837-846.

Vrijhoef, R. (1998). *Co-makership in construction: towards construction supply chain management.* MSc thesis. Delft University of Technology / VTT Building Technology, Espoo. 180 p.

Vrijhoef, R., Cuperus, Y. and Voordijk, J.T. (2002). "Exploring the connection between open building and lean construction: defining a postponement strategy for supply chain management". In: Formoso, C.T. (ed.). *Proceedings 10th Annual IGLC Conference.* 6 August 2002, UFRGS, Gramado, Brazil.

Vrijhoef, R. and De Ridder, H.A.J. (2005). "Supply chain integration for achieving best value for construction clients: client-driven versus supplier-driven integration". In: Sidwell, A.C. and Betts, M. (eds.). *Proceedings QUT Research Week.* 4 July 2005, Brisbane. QUT, Brisbane. Proceedings on CD-Rom.

Vrijhoef, R. and Koskela, L. (2000). "The four roles of supply chain management in construction". *European Journal of Purchasing & Supply Management,* 6 (3-4), 169-178.

Vrijhoef, R. and Koskela, L. (2005a). "A critical review of construction as a project-based industry: identifying paths towards a project-independent approach to construction". In: Kazi, A.S. (ed.). *Proceedings CIB Combining Forces.* Volume 9: Learning from Experience: New Challenges, Theories and Practices in Construction. 13 June 2005, Helsinki. VTT, Espoo. 13-24.

Vrijhoef, R. and Koskela, L. (2005b). "Revisiting the three peculiarities of production in construction". In: Kenley, R. (ed.). *Proceedings 13th Annual Lean Construction Conference (IGLC-13).* 19 July 2005, Sydney. 155-163.

Vrijhoef, R. and Koskela, L. (2005c). "Structural and contextual comparison of construction to other project-based industries". In: Ruddock, L. and others (eds.). *Proceedings IPRC 2005*. Volume 2. 14 April 2005, Salford. University of Salford, Salford. 537-548.

Vrijhoef, R., Koskela, L. and Howell, G.A. (2001). "Understanding construction supply chains: an alternative interpretation". In: *Proceedings 9th Annual IGLC Conference*. 6 August 2001, Singapore. 15 p.

Vrijhoef, R. and Voordijk, J.T. (2004). "Supply chain implications of platform strategies in construction". In: Ruddock, L. (ed.). *Proceedings 4th International Postgraduate Research Conference in the Built and Human Environment (IPRC)*. 1 April 2004, University of Salford, Salford. 77-87.

Wacker, J.G. (1998). "A definition of theory: research guidelines for different theory-building research methods in operations management". *Journal of Operations Management,* 16, 361-385.

Waller, B. (2004). "Market responsive manufacturing for the automotive supply chain". *Journal of Manufacturing Technology Management,* 15 (1), 10-19.

Warburton, R.D.H. and Stratton, R. (2002). "Questioning the relentless shift to offshore manufacturing". *Supply Chain Management: An international Journal,* 7 (2), 101-108.

Warszaswki, A. (1990). *Industrialisation and robotics in building: a managerial approach*. Harper & Row, New York.

Watanabe, C. and Ane, B.K. (2004). "Constructing a virtuous cycle of manufacturing agility: concurrent roles of modularity in improving agility and reducing lead time". *Technovation,* 24 (7), 573-583.

Webster, M., Alder, C. and Muhlemann, A.P. (1997). "Subcontracting within the supply chain for electronics assembly manufacture". *International Journal of Operations & Production Management,* 17 (9), 827-841.

Wegelius-Lehtonen, T. and Pahkala, S. (1998). "Developing material delivery processes in cooperation: an application example of the construction industry". *International Journal of Production Economics,* 56/57, 689-698.

Weick, K.E. (1989). "Theory construction as disciplined imagination". *Academy of Management Review,* 14 (4), 516-531.

Wheelwright, S.C. and Clark, K.B. (1992). *Revolutionizing product development: quantum leaps in speed efficiency and quality*. The Free Press, New York.

Williams, T., Maull, R. and Ellis, B. (2002). "Demand chain management theory: constraints and development from global aerospace supply webs". *Journal of Operations Management,* 20, 691-706.

Williamson, O.E. (1975). *Markets and hierarchies: analysis and antitrust implications.* Free Press, New York.

Williamson, O.E. (1985a). *New institutional economics.* Free Press, New York.

Williamson, O.E. (1985b). *The economic institutions of capitalism: firms, markets, relational contracting.* Free Press, New York.

Williamson, O.E. (1991). "Comparative economic organization: the analysis of discrete structural alternatives". *Administrative science quarterly,* 36 (2), 269-296.

Williamson, O.E. (2008). "Outsourcing: transaction cost economics and supply chain management". *Journal of Supply Chain Management,* 44 (2), 5-16.

Williamson, O.E. and Winter, S. (1991). *The nature of the firm.* Oxford University Press, New York.

Winch, G.M. (1989). "The construction firm and the construction project: a transaction cost approach". *Construction Management and Economics,* 7, 331-345.

Winch, G.M. (2001). "Governing the project process: a conceptual framework". *Construction Management and Economics,* 19, 799-808.

Winch, G.M. (2003). "Models of manufacturing construction process: the genesis of reengineering construction". *Building Research & Information,* 31 (2), 107-118.

Winch, G.M. (2006). "Towards a theory of construction as production by projects". *Building Research & Information,* 34 (2), 164-174.

Winch, G.M. and Champagnac, E. (1995). "The organisation of building projects: an Anglo-French comparison". *Construction Management and Economics,* 13, 3-14.

Winfield, I. and Hay, A. (1997). "Toyota's supply chain: changing employee relations". *Employee Relations,* 19 (5), 457-465.

Winograd, T. and Flores, F. (1986). *Understanding computers and cognition: a new foundation for design.* Ablex, Norwood.

Wislocki, P. (2001). "Cepezed: case studies in designer-led integrated procurement". *Journal of Architecural and Planning Research,* 18 (4), 271-285.

Wolters, M.J. (2000). *The business of modularity and the modularity of business*. Dissertation. Erasmus Research Institute of Management, Rotterdam.

Womack, J.P. and Jones, D.T. (1994). "From lean production to the lean enterprise". *Harvard Business Review,* March-April, 93-103.

Womack, J.P. and Jones, D.T. (1996). *Lean thinking*. Simon & Schuster, New York.

Womack, J.P., Jones, D.T. and Roos, D. (1990). *The machine that changed the world*. Rawson Associates, New York. 323 p.

Worren, N., Moore, K. and Cardona, P. (2002). "Modularity, strategic flexibility, and firm performance: a study of the home appliance industry". *Strategic Management Journal,* 23 (12), 1123-1140.

Wortmann, J.C. (1992). "Production management systems for one-of-a-kind products". *Computers in Industry,* 19 (1), 79-88.

Woudhuysen, J. and Abley, I. (2004). *Why is construction so backward?* Wiley, Chichester. 336 p.

Xie, S.Q., Tu, Y.L., Fung, R.Y.K. and Zhou, Z.D. (2003). "Rapid one-of-a-kind product development via the internet: a literature review of the state-of-the-art and a proposed platform". *International Journal of Production Research,* 41 (18), 4257-4298.

Yen, B.P.C. (2002). "Electronic commerce front-end in apparel supply chain". *Computers & Industrial Engineering,* 42, 471-480.

Yeung, J.F.Y., Chan, A.P.C. and Chan, D.W.M. (2007). "The definition of alliancing in construction as a Wittgenstein family resemblance concept". *International Journal of Project Management,* 25 (2007), 219-231.

Yigit, A.S. and Allahverdi, A. (2003). "Optimal selection of module instances for modular products in reconfigurable manufacturing". *International Journal of Production Research,* 41, 4063-4074.

Yin, R.K. (1989). *Case study research: design and methods*. Sage Publications, Newbury Park CA.

Yu, Z., Yan, H. and Cheng, T.C.E. (2001). "Benefits of information sharing with supply chain partnerships". *Industrial Management & Data Systems,* 101 (3), 114-121.

Zailani, S. and Rajagopal, P. (2005). "Supply chain integration and performance: US versus East Asian companies". *Supply Chain Management: An international Journal,* 10 (5), 379-393.

Appendices

Case Study Protocol

The case study protocol is a tactic to ensure the reliability of the case study research and to guide the researcher (Yin 1989). This protocol contains the procedures followed so they could be replicated. The protocol distinguishes between the manufacturing cases and the building cases, however the former have not been conducted as genuine case studies. For each of the groups of cases the procedures have been listed chronologically below. This includes the procedures of the studies of the manufacturing cases as well as the building case studies reported in Chapters 6 and 7 respectively. The protocol excludes the further case study analysis in Chapter 8. An overview of the broader approach to the case studies and the methods used following Eisenhardt (1989) have been described in Chapter 2.

Manufacturing cases

The procedures followed in the manufacturing cases were aimed at studying six examples of supply chain integration as applied in six different manufacturing industries by using secondary data. The procedures were as follows:

- Potential cases had been listed on a long list.
- Industry typology had been leading for the selection of the cases.
- The selected cases had been explored.
- A data collection plan had been made including a list of potential data sources.
- Data sources had been assessed whether they could provide data.
- Secondary data had been gathered from the selected data sources by desk research.
- Additional non-selected data sources and incidental data findings had been recorded.
- Data collected had been documented and stored.
- Data had been analysed by narrative analysis.
- Open coding had led to first order themes.
- Axial coding had led to second order themes.
- Second order themes had been used for the further case study analysis of the manufacturing as well as the building cases.

Building cases

The procedures followed in the building cases were aimed at studying nine applications of supply chain integration by different firm types in the building supply chain from clients to suppliers by using a combination of primary and secondary data. The procedures were as follows:

- Potential cases had been listed on a long list.
- Firm typology had been leading for the selection of the cases.
- The selected cases had been explored.
- A data collection plan had been made including a list of potential data sources.
- Data sources had been assessed whether they could provide data.
- Secondary data had been gathered from the selected data sources by desk research.
- Data collected had been documented and stored.
- Data had been analysed by narrative analysis.
- Interviews had been prepared based on secondary data collected.
- Interviews had been semi-structured following the analytical model.
- Multiple interviews had been taken per case study.
- Interviews had been transcribed.
- Transcriptions had been analysed manually and by use of qualitative data analysis software.
- Second order themes from the manufacturing cases had been used for the analysis.
- Case studies had been presented in reports, quotes and thematic connections.

Case study topics

The manufacturing cases as well as the building cases had been organised around the subjects given by the analytical model presented in Chapter 4. The ten factors of analysis of the model represented the directions for both searching for secondary data, e.g. documentations, media reports etc., and for the questions in the semi-structured interviews in the building cases. Per factor of analysis the main questions were: what had been done in the particular case on this subject, how had this been done, and where had it led to.

Case study reports

The case study reports typically included overviews of the data retrieved in each case, summaries of the secondary data found in each case, and narratives and transcripts of the interviews in the building cases. The case study reports had been organised following the subjects given by the analytical model, i.e. the ten factors of analysis presented in Chapter 4. The draft texts of the case study reports had been the textual basis leading to the case study descriptions presented in Chapters 6 and 7.

Themes retrieved from coding

Table 69: Overview of 1st and 2nd order themes based on open and axial coding of manufacturing cases

1st order themes	2nd order themes
Retrieved by open coding of the manufacturing cases	Derived by axial coding of 1st order themes into larger 2nd order themes
Binding suppliers for a long term Absolute supplier commitment to manufacturer Manufacturer leads product development Manufacturer controls suppliers as external business units	Focal firm dominance
Shortening lead times Shortening time to market Retrieving customer and market information Considering the end customer	Market responsiveness
Process flow Highly developed and tested standard procedures Fast and faultless logistics Non-acceptance of failures Joint understanding of what excellence is	Operational excellence
Pushing the market by constantly new products Seducing clients to buy new products Developing future technology to install in new products	Market push
Cost reductions Increasing profitability Sharing risks and benefits Investments must be profitable Managing for result	Commercial focus
Managing for speed Short product life cycles Time lapse on product development Time to market must be short to beat the competition	Speed
Supplier complying to manufacturers processes Manufacturer controls supplier processes Control and monitoring mechanisms for integrated process Advanced ICT systems for process support	Absolute control
Manufacturers constantly monitoring the supply chain Constant up-to-date information for everyone Information is accurate and leading operations	Real-time information
Low inventories, just in time Preventing oversupply nor stock-outs Fast obsolescence of products Reduction of capital retained in semi-products	Inventory control
Being in the technological forefront Manufacturer as dominant product specialist Suppliers as intelligent system/parts specialists	Specialist roles

Constant quality focus Suppliers complying to high quality Non-acceptance of quality problems Direct action in case of quality problems	Total quality
Designing products for interchangability, modularity of parts Designing products for manufacturability Postponing customisation of products	Design rationality
Long-term cooperation Reducing numbers of suppliers, focusing on key suppliers Key suppliers involved in key business functions Joint product development Commitment to mutual prosperity	Strategic cooperation
Being an innovator Emphasis on technological advancement R&D to sustain technological leadership	Technological supremacy
Open information and knowledge sharing Personnel exchange Diffusing corporate boundaries Commitment to joint undertaking	Teamwork

Data gathering in the manufacturing cases

The secondary data gathered in the manufacturing cases were publicly available, and displayed below.

Table 70: Overview of data gathered in the manufacturing cases

Cases	Literature, reports	Additional publications	Internet sources
Automotive: Toyota	• Toyota's supply chain: changing employee relations, by I. Winfield and A. Hay (1997) • Creating and managing a high performance knowledge-sharing network: the Toyota case, by J.H. Dyer and K. Nobeoka (Dyer & Nobeoka 2000) • Toyota supply chain management: a strategic approach to the principles of Toyota's renowned system, by A. Iyer et al. (2009)	• Shorter is better for Toyota's supply chain: automaker aims to localize production and supply base, by D. Hannon, in Purchasing, 14 August 2008 • The Toyota Way and Supply Chain Management, by Jeffrey K. Liker	• www.toyota.com • www.suite101.com/content/supply-chain-management-in-ford-toyota-and-hp-a289041 • www.scribd.com/doc/13754548/Supply-Chain-Management-Of-Honda-Toyota
Aerospace: Airbus	• A case study of the Airbus West Factory Broughton North Wales, by A. Fleming (unpublished) • Design of a supply chain quality logistics model based on the production system of Airbus Deutschland GmbH, by T. Duivenvoorden	• Comparative Analysis of Supply Chain Management Practices by Boeing and Airbus: Long-Term Strategic Implications, by T.C. Horng and K. Bozdogan • Robotboot voor A380, in De Ingenieur 20-2-2009	• www.rfidnews.org/2008/04/25/airbus-supply-chain-visualized • www.indiadaily.com/editorial/13367.asp • www.smalallah.com/2010/06/19/airbus-supply-chain-management/ • www.icmrindia.org/casestudies/catalogue/Operations/OPER092.htm
Computers: Dell	• The power of virtual integration: an interview with Dell Computer's Michael Dell, by J. Magretta (1998a)	• Implementation Study: Dell IT Scales Supply Chain Management with Oracle RAC 10g, by Dave Jaffe et al.	• www.dell.com/content/topics/global.aspx/about_dell/values/supp_citizen/supply?c=us&l=en • supply-chain-case-studies.blogspot.com/2007/11/dell-supply-chain-management-case-study.html • www.scdigest.com/assets/First • Thoughts/08-04-10.php • www.icmrindia.org/casestudies/catalogue/Operations/OPER063.htm

Electronics: ASML	• Emergent Supply Networks: System Dynamics Simulation of Adaptive Supply Agents, by H. Akkermans (2001) • Platform-Driven Development of Product Families: Linking Theory with Practice, by J.I. M. Halman et al (2003) • New SCOP Method in ASML Supply Chain: Application of Enhanced Synchronized Base Stock in planning of Supply Chain Planning environment, by W.S. De Jong (2010)	• N/A	• www.asml.nl
Clothing/ textile: Zara	• How Zara fashions its supply chain (2005)	• Retail @ the speed of fashion, by D. Dutta	• hbswk.hbs.edu/archive/4652.html • www.icmrindia.org/casestudies/catalogue/Operations/OPER055.htm • ivythesis.typepad.com/term_paper_topics/2009/08/case-study-zaras-vertical-supply-chain.html • www.delaatstemeter.nl/slimme-laatste-meters/zara-het-geheim-van-de-supersnelle-supply-chain/
Food/ rocery: Wal-mart	Wall-mart's supply chain management practices, by P.M. Chandran (2003)	N/A	• mohanchandran.files.wordpress.com/2008/01/wal-mart.pdf • www.casestudyinc.com/case-study-walmart-supply-chain • www.foodproductiondaily.com/Supply-Chain/Wal-Mart-begins-supply-chain-revolution • hbswk.hbs.edu/archive/3616.html • www.asaresearch.com/ecommerce/supplychain.htm

Data gathering in the building cases

The primary data gathered in the building cases had been retrieved directly from the cases. The secondary data were publicly available. The data have been made anonymous below. The data have been stored in the case study database.

Table 71: Overview of data gathered in the building cases

Cases	Primary data		Secondary data
Firms	Interviews	Internal documentation	External publications
Client 1 Contractor Specialist contractor Supplier	Manager Director Project manager Director Director	• Annual report • Evaluation of projects based on supply chain integration • Evaluation of project procurement • Strategic plan for implementation of supply chain integration • Project plan for implementation of supply chain integration • Internal newsletters	• External newsletters External report • BSc final work of student on the subject • MSc final work of student on the subject • Company websites • Paper articles • Magazine articles • Trade journal articles
Client 2 Two contractors	Manager Director Project manager	• Annual report • Project planning for implementation of supply chain integration on projects • Evaluation of projects following supply chain integration	• MBA final work of student on the subject • Company websites • Paper articles • Magazine articles • Trade journal articles
Developer	Managing director Project manager	• Annual report • Company presentations • Project data	• External presentations • Company websites • Paper articles • Magazine articles • Trade journal articles
Designer 1	Director Project manager	• Annual report • Project data	• Scientific paper analysing the supply chain approach • Project portfolio • Company websites • Paper articles • Magazine articles • Trade journal articles

Designer 2	Managing director Manager Project manager Internal developer/ contractor	• Annual report • Project data • Process maps	• Project portfolio • Company websites • Paper articles • Magazine articles • Tradc journal articles
Builder 1 Supplier	Managing director Project manager Manager	• Annual report • Project data • Evaluation report of business model	• External presentations • Company websites • Paper articles • Magazine articles • Trade journal articles
Builder 2 Architect Supplier	Manager Owner/director Manager	• Annual report • Project data • Product information • Building Information Modelling process maps and protocols	• External presentation • Press releases • External newsletters • Sales brochures • Company websites • Paper articles • Magazine articles • Trade journal articles
Builder 3 Architect Supplier	Managing director Technical manager Project manager Owner/director Managing director	• Annual report • Project data • Product information • R&D reports	• External newsletters • Sales brochures • Company websites • Paper articles • Magazine articles • Trade journal articles
Suppliers Installation firm Contractor	Manager R&D manager Manager Purchaser/ developer	• Annual report • Project data • Product information • R&D reports	• Sales brochures • Company websites • Paper articles • Magazine articles • Trade journal articles

Glossary of Abbrevations and Terms

The below abbreviations and terms have been explained and defined as follows within the scope of this thesis.

Alignment	Adjusting the relative positions of firms in the supply chain for better and closer mutual connection
AR	Accurate Response
BIM	Building Information Model
Centralisation	Stressing the centrality and controlling the supply chain from a central position
Coherence	Systematic connection between firms in the supply chain
Collectivism	The idea of belonging to a group and the willingness to act together as a group
Continuity	Continuous duration particularly of work, beyond the temporal scope of projects
Core competencies	Essential capabilities of organisations characterising their identity
CPFR	Collaborative Planning, Forecasting and Replenishment
Cultural alignment	Adjusting respective business cultures to increase commitment and channel behaviour supporting and serving the supply chain as it were a 'single firm'
DBFMO	Design-Build-Finance-Maintain-Operate
DC	Distribution Centre
Demand chain integrator	The central organisation or firm in the demand chain bringing all demand together and often the focal point representing the collective demand
ECR	Efficient Customer Response
ERP	Enterprise Resource Planning
Extended enterprise	The idea of the supply chain as an enterprise encompassing the firms constituting the supply chain
Extended delivery process	The sum of all processes throughout the supply chain jointly producing and delivering the end product to the end customer
Flow of production	The idea that production progresses smoothly through a seamless production channel
Focal firm	The central organisation or firm in the supply chain
Fragmentation	The effect of the supply chain or production system not being a coherent whole, but rather consisting of many different activities and many different firms
GOO	Global Operations Organisation

Human resource management	Joint approach to personnel, including joint staffing and training, fostering integrated working and stability of the workforce throughout the supply chain
ICT	Information and Communication Technology
Information exchange	Firms in the supply chain share information and knowledge in a joint information system
Integration of business activities	Internalising business activities, and additionally controlling external business activities as well
Integration of operations and processes	Connected governance of processes and operations to achieve flow, and increase the effectiveness and efficiency
Interdependency	The effect of causality among activities performed in the supply chain
Internalisation	Acquiring or absorbing external activities within the own corporate boundaries
ISO	International Standardisation Organisation
JIT	Just In Time
Market approach and marketing	Collaborative market approaches with the aim of delivering integrated products to the end market effectively and efficiently
Modularity	Products or production consisting of interchangeable elements that can be produced separately
MRP	Manufacturing Resource Planning
Partner sourcing and collaboration strategies	Establishing strategic agreements and long-term collaborative arrangements with key partners in the supply chain
PDA	Personal Digital Assistant
PFI	Privately Financed Initiative
Planning and logistics control	Jointly planning operations and processes, and jointly organise and manage logistics
Platform strategy	Product and production strategy promoting modularity, interchangeability and standardisation of parts
Postponement	The attempt to postpone the moment the final product is assembled or customised for the end customer
Predictability	The level to which production in particular can be predicted
Product development and design	Developing and designing products jointly and rationally applying product and production modularisation
QMS	Quality Management System
QR	Quick Response
Quasi-firm	The general contractor and specialist subcontractors forming a stable organisational unit when conditions permit (Eccles 1981)
Quasi-integration	An hybrid form of integration between market and hierarchy
R&D	Research and Development
Relational contracting	Contracting based upon a relationship of trust between parties

Repetitiveness	The level of repeated application of products, processes and organisational arrangements
Responsiveness	The reaction rate to which one in particular is able to respond to demand or market developments
RFID	Radio Frequency Identification
SCM	See: Supply chain management
Segmentation	The level to which a population or sector is divided in different groups or divided in general
SKU	Stock Keeping Unit
SME	Small and Medium Enterprise
Socialisation	Establishing on a social basis among firms or individuals
Strategic collaboration	Long-term and intensified collaboration between firms on a strategic level
Supply chain	Network of firms interacting to deliver a product or service to the end customer, linking flows from raw material supply to final delivery
Supply chain collaboration, cooperation	Establishment of collaborative systems and arrangements among firms in the supply chain. See also table below for definitions found in literature
Supply chain integration	Arrangement in supply chains referring to higher levels of integration, notably integration of shared information and integrated management of supply chain activities and operations including product development, materials manufacture and product assembly. See also table below for definitions found in literature
Supply chain integrator	The central organisation or firm in the supply chain integrating and coordinating activities and processes throughout the supply chain, often referred to as the focal firm
Supply chain management	Management focussing on inventory planning and logistics management towards more comprehensive outsourcing strategies for instance including economic issues and risk sharing with suppliers. See also table below for definitions found in literature
Supply chain partnering, alliances	Fostering equal relations between supply chain firms viewed as partners. Often such alliances and partnerships reside on a strategic or tactical level between the firms involved, for instance including integrated arrangements to finance and risk sharing. See also table below for definitions found in literature
TCE	Transaction Cost Economics
Togetherness	The idea of being closely connected
Total quality	Continuously improving the integrated quality of products and processes throughout the supply chain
Total quality management	Managing and achieving total quality to processes across corporate boundaries
TPS	Toyota Production System

TQM	See: Total quality management
Transparency	The characteristic that one can see through a system such as the supply chain
Vertical integration	Internalisation of neighbouring activities in the supply chain
Virtual organisation	An imaginary organisation consisting of multiple actual organisations
VMI	Vendor Managed Inventory

Table 72: Definitions of supply chain integration and related concepts found in literature

Definitions	References
Supply chain integration	
Supply chain integration is only a natural result of redesigned business processes not realignment of existing functional organisations.	Hewitt (1992)
(..) a series of suppliers, manufacturers, and customers link together into seamlessly integrated supply chains.	Frohlich and Westbrook (2001)
(…) synchronise interdependent activities, ensure visibility to match supply and demand, align actions and decision with the chain profitability, and acquire new capabilities from joint efforts.	Simatupang et al. (2002)
(…) closer contact with customers and a greater dissemination of information, both of which are seen as indicators of greater integration of a value chain.	Armistead and Mapes (1993)
(…) implementing supply chain concepts or principles in a much more effective and integrated way, including information sharing, multi-party collaboration, design for supply chain management, postponement for mass customization, outsourcing and partnerships, and extended or joint performance measures.	Lee and Whang (1999)
(…) forward and backward integration with valued suppliers and customers (…).	Fawcett and Magnan (2002)
Supply chain collaboration	
(...) collaboration with customers and suppliers (…) facilitating behavioral change, that is, the shifting from traditional arms-length or even adversarial attitudes to a partnership perspective that fosters cooperation and a freer exchange of information.	Stank et al. (2001)
(…)jointly creating the common pace of information sharing, replenishment, and supply synchronization in the system reduces both excess inventory and is essential to avoid the costly bullwhip effect (…).	Holweg et al. (2005)
Collaboration where two or more parties in the supply chain jointly plan a number of promotional activities and work out synchronised forecasts, on the basis of which the production and replenishment processes are determined	Skjoett-Larsen et al. (2003)
(…) collaboration among all participants in the value chain, whatever their size, function, or relative position.	Sahay (2003)
(…) supply chain approach limiting collaboration to a small but potentially critical number of customers and suppliers (..).	Barratt (2004)

Supply chain partnership	
(...) direct, long term association, encouraging mutual planning and problem solving efforts.	Maloni and Benton (1997)
(...) information sharing-based partnership giving incentives to guarantee supply reliability and induce cooperation.	Yu et al. (2001)
(...) mutual cooperation and coordination among partners to decide the retail and the wholesale price, profit margins and inventory level in stock in order to get a greater market share and hence higher revenues.	Chauhan and Proth (2005)
(...)closely integrated, mutually beneficial relationships that enhance supply chain performance	Lambert et al. (2004)
Supply chain management	
Supply chain management focuses attention on the interactions of channel members to produce an end product/service that will provide value for the end user. (...) the entire sourcing, value-added and marketing activities of the overall link of firm up to final customers. Supply chain management crystallises (...) by providing a process framework that enables firms to engage in co-evolvement rather than competition.	Bechtel and Yayaram (1997)
(...) managing value across organisational boundaries, which mandates new forms and new ways of managing inter-corporate relationships.	Caldwell and Down (1997)
The supply chain concept consists of actively managed channels of procurement and distribution. It is the group of firms that add value along product flow from original raw materials to final customers. It concentrates on relational factors rather than transactional ones.	Cavinato (1992)
Supply chain management covers the flow of goods from supplier through manufacturing and distribution chains to the end user. Supply chain management is the management of upstream and downstream relationships with suppliers and customers to deliver superior customer value at less cost to the supply chain as a whole.	Christopher (1992)
Supply chain management is viewed as lying between fully-vertically-integrated systems and those where each channel member operates completely independently. Supply chain management is an approach whereby the entire network from which suppliers through the ultimate customer, is analysed and managed in order to achieve the "best" outcome for the whole system.	Cooper and Ellram (1993)
Supply chain management is an approach whereby the entire network from which suppliers through the ultimate customer, is analysed and managed in order to achieve the "best" outcome for the whole system.	Cooper and Gardner (1993)
Supply chain management is an integrative approach to using information to manage inventory throughout the channel, from source of supply to end-user. The integration of all key business processes across the supply chain is what we are calling supply chain management.	Cooper et al. (1997)
(...) approach (that) will enable a manufacturing operation to better manage its supply chain, ultimately improving customer satisfaction levels while reducing overall costs.	Davis (1993)

Supply chain management is defined as an integrative approach to dealing with the planning and control of the materials flow from suppliers to end-users. Supply chain management is an integrative approach to using information to manage inventory throughout the channel, from source of supply to end-user.	Ellram (1990)
Supply chain management is defined as an integrative approach to dealing with the planning and control of the materials flow from suppliers to end-users. (...) an integrative philosophy to manage the total flow of a distribution channel from the supplier to the ultimate user.	Ellram and Cooper (1990)
Supply chain management has been referred to as an incentive to establish partnerships between parties in the supply chain. In its broadest context supply chain management is a strategic tool used to enhance overall customer satisfaction that is intended to improve a firm's competitiveness and profitability.	Giunipero et al. (1996)
Supply chain management is the management of a network of interconnected businesses involved in the ultimate provision of product and service packages required by end customers. It spans all movement and storage of raw materials, work-in-process inventory, and finished goods from point of origin to point of consumption.	Harland (1996)
Supply chain management is really an operations approach to procurement. It requires all participants of the supply chain to be properly informed. With supply chain management, the linkage and information flow between various members of the supply chain are critical to overall performance.	Håkansson and Johannson (1994)
Supply chain management covers the flow of goods from supplier through manufacturer and distributor to the end user.	Houlihan (1988)
Supply chain management is really an operations approach to procurement. It requires all participants of the supply chain to be properly informed. With supply chain management, the linkage and information flow between various members of the supply chain are critical to overall performance.	Johannson (1994)
Supply chain management deals with the total flow of materials from suppliers through end users.	Jones et al. (1985)
Supply chain management is the integration of key business processes across the supply chain for the purpose of adding value for customers and stakeholders.	Lambert (2008)
Supply chain management is the integration of key business processes from end user through original suppliers that provide products, services and information that add value for customers and other stakeholders.	Lambert et al (1998)
One if the defining features of supply chain management is the extent of vision required on the part of managers with respect to their value-adding system, and the influence they must seek to have over it.	Lamming (1996)
(...) a networking approach to value chain optimisation.	Lamming (2005)
Supply chain management is the systemic, strategic coordination of the traditional business functions and the tactics across these business functions within a particular company and across businesses within the supply chain, for the purposes of improving the long-term performance of the individual companies and the supply chain as a whole.	Mentzer et al. (2001)

Supply chain management covers the flow of goods from the supplier through the manufacturer and distributor to the end user.	Novack and Simco (1991)
(...) the concept of supply chain management is centred on organisational restructuring and extends to the development of a company-wide collaborative culture but also embraces a strong sense of the integration of all activities which control the timing and synchronisation of material flows.	Rich and Hines (1997)
(...) supply chain management is used to refer to the chain linking each element of the production and supply process from raw materials through to the end customer.	Scott and Westbrook (1991)
Control the flow of material from suppliers, through the value adding (production) processes and distribution channels, to customers.	Stevens (1990)
(...) technique that looks at all the links in the chain from raw materials suppliers through various levels of manufacturing to warehousing and distribution to the final customer.	Turner (1993)

List of Figures

Figure 1: Three characteristics of building on product, production and industry level. Adapted from (Vrijhoef & Koskela 2005b) 2

Figure 2: Characteristics of building and their causal relations (Vrijhoef & Koskela 2005b) 3

Figure 3: Schematic representation of the demand and supply system around a built object (Vrijhoef & De Ridder 2005) 4

Figure 4: Different delivery types in the building supply chain form make-to-order to concept-to-order (Bossink & Vrijhoef 2009) 5

Figure 5: Production system types varying from (one-off) designing to (repetitive) making. The production system of building is deemed to be engineer-to-order (Adapted from Ballard 2005) 5

Figure 6: Changes in firm size in the Dutch building industry 1998-2006. Data provided by Bouwend Nederland on www.bouwendnederland.nl 6

Figure 7: Problems progressing through the building supply chain (Vrijhoef 1998) 7

Figure 8: The idea of the supply chain integrator in building tying together and coordinating different flows through the supply chain. Adapted from (Segerstedt & Olofsson 2010, Vrijhoef & De Ridder 2005) 8

Figure 9: Structure of the thesis 16

Figure 10: Deduction and induction as applied in this research. Adapted from Love et al. (2002a) 23

Figure 11: Multiple case study approach to theory building as applied in this thesis. Adapted from Yin (1989) and Eisenhardt (1989) 24

Figure 12: Porter's value chain indicating the perspectives on the supply chain. Adapted from Porter (1985) 55

Figure 13: Four theoretical perspectives on the building supply chain. Adapted from Vrijhoef et al. (2003) 57

Figure 14: Generic configuration of a supply chain in manufacturing (Vrijhoef & Koskela 2000) 59

Figure 15: Decoupling point dictating the market strategy in the supply chain (Hoekstra & Romme 1992) 76

Figure 16: Supply chain management maturity model by the Supply Chain Council (Lockamy & McCormack 2004) 81

Figure 17: Levels of supply chain integration 81

Figure 18: Conceptual differences between process and discrete manufacturing (Fransoo & Rutten 1994) 92

Figure 19: Type 1 supply chain structure: early differentiation, convergent assembly (Lin & Shaw 1998) 93

Figure 20: Type 2 supply chain structure: delayed differentiation, divergent assembly (Lin & Shaw 1998) 94

Figure 21: Type 3 supply chain structure: responsiveness, divergent differentiation (Lin & Shaw 1998) 94

Figure 22: Overview of the building case studies 133

List of Tables

Table 1: Initial direction of the research objective: topics of supply chain integration in theory, applications in manufacturing and direction in building. 12

Table 2: Process of theory building from case study research in this thesis (Adapted from Eisenhardt 1989). 25

Table 3: Overview of the thesis and research structure. 29

Table 4: Characteristic differences between building and other industries (Vrijhoef & Koskela 2005c) 40

Table 5: Differentiating between demand and supply perspectives on the role of the system integrator (Vrijhoef & Koskela 2005a) 43

Table 6: Overview of contributing topics to the concept of building supply chain integration 45

Table 7: Overview of constructs presented in this chapter, their conceptual contributions and gaps 50

Table 8: Support activities included in the thesis based on Porter's value chain model (Porter 1985) 56

Table 9: Theoretical framework for supply chain integration in building 66

Table 10: Overview of constructs presented in this chapter, their conceptual contributions, and gaps 67

Table 11: Directions for the operationalisation of the theoretical framework 72

Table 12: Factors of analysis for supply chain integration 79

Table 13: Analytical model for supply chain integration 82

Table 14: Properties of the three types of supply chain. Adapted from (Lin & Shaw 1998) 95

Table 15: Manufacturing industries and cases selected for this research 96

Table 16: Overview of applications of supply chain integration in the automotive industry, notably Toyota 100

Table 17: Overview of applications of supply chain integration in the aerospace industry, notably Airbus 104

Table 18: Overview of applications of supply chain integration in the computer industry, notably Dell 109

Table 19: Overview of applications of supply chain integration in the electronics industry, notably ASML 114

Table 20: Overview of applications of supply chain integration in the clothing and textile industry, notably Zara 120

Table 21: Overview of applications of supply chain integration in the food and grocery sector, notably Wal-Mart 125

Table 22: Summary of applications of supply chain integration found in manufacturing per supply chain type 126

Table 23: Case overview of the housing corporation in operational cooperation with builders (Client 1) 139

Table 24: Case overview of the housing corporation in legal partnership with builders (Client 2) 144

Table 25: Case overview of the full-service project developer (Developer) 148

Table 26: Case overview of the developing and contracting architect (Designer 1) 152

Table 27: Case overview of the architectural engineer (Designer 2) 156

Table 28: Case overview of the catalogue builder (Builder 1) 159

Table 29: Case overview of the ICT-driven builder (Builder 2) 164

Table 30: Case overview of the lean-driven builder (Builder 3) 168

Table 31: Case overview of the three suppliers integrating products and site installation (Suppliers) 172

Table 32: Summary of applications of supply chain integration found in the building cases 176

Table 33: Within-case analysis of the case of the housing corporation in operational cooperation with builders (Client 1) 181

Table 34: Within-case analysis of the case of the housing corporation in legal partnership with builders (Client 2) 182

Table 35: Within-case analysis of the case of the full-service project developer (Developer) 183

Table 36: Within-case analysis of the case of the developing and contracting architect (Designer 1) 184

Table 37: Within-case analysis of the case of the architectural engineer (Designer 2) 185

Table 38: Within-case analysis of the case of the catalogue builder (Builder 1) 186

Table 39: Within-case analysis of the case of the ICT-driven builder (Builder 2) 187

Table 40: Within-case analysis of the case of the lean-driven builder (Builder 3) 188

Table 41: Within-case analysis of the case of the three suppliers integrating products and site installation (Suppliers) 189

Table 42: Similarities and differences between the findings from the building case studies based on 2nd order themes 191

Table 43: Development levels of the applications of supply chain integration found in the building case studies 194

Table 44: Similarities and differences between the findings from the manufacturing comparator studies based on 2nd order themes 198

Table 45: Development levels of the applications of supply chain integration found in the manufacturing comparator studies 199

Table 46: Cross-case analysis between the main findings from the building case studies and the manufacturing comparator studies 202

Table 47: Reflecting on the importance of the ten factors for supply chain integration in the building industry versus manufacturing 204

Table 48: Factor 1: Repetitiveness 210

Table 49: Factor 2: Integration of business activities 211

Table 50: Factor 3: Partner sourcing and collaboration strategies 212

Table 51: Factor 4: Integration of operations and processes 213

Table 52: Factor 5: Planning and logistics control 214

Table 53: Factor 6: Quality management 215

Table 54: Factor 7: Information exchange 216

Table 55: Factor 8: Product development and design 217

Table 56: Factor 9: Market approach and marketing 218

Table 57: Factor 10: Cultural alignment 219

Table 58: Factor 11: Human resource management 220

Table 59: Categorising the factors by relations to goals 221

Table 60: Relating the factors 222

Table 61: Clustering the factors 224

Table 62: Identifying main goals based on the clusters of factors 225

Table 63: Relating the goals 226

Table 64: Clustering the goals 227

Table 65: Clusters of factors 228

Table 66: Clusters of goals 229

Table 67: Formulation of hypotheses, connections to theory and manufacturing, and applicability to building 230

Table 68: Specification of the concept of supply chain integration per firm type 238

Table 69: Overview of 1st and 2nd order themes based on open and axial coding of manufacturing cases 284

Table 70: Overview of data gathered in the manufacturing cases 286

Table 71: Overview of data gathered in the building cases 288

Table 72: Definitions of supply chain integration and related concepts found in literature 293

Summary

Supply chain integration in the building industry: the emergence of integrated and repetitive strategies in a fragmented and project-driven industry

The building industry has been criticised extensively for being wasteful, not innovative and unproductive. In those criticisms, reference is often made to other sectors of industry, particularly manufacturing, that functions and produces its products in more effective and efficient ways than the building industry does apparently. One of the major differences observed between building and manufacturing, among other things, has been the organisation and coordination of the supply chain. The organisation and coordination of supply chains have been observed and conceptualised in different ways in building and manufacturing based on the specifically characteristic differences between those sectors. In building, the organisational approaches of supply chains have particularly been influenced by the one-off, temporal nature of projects; the large number of firms involved in the definition, design, manufacture and assembly of built objects involving many relatively small firms; the dispersed power and governance regimes; and the initiating role of clients. The high levels of segmentation and the low levels of repetition in the building industry have led to many problems and deficiencies in the production system. In the building supply chain, the two phenomena have led to typical problems of lack of control and decreasing performance, which tend to reinforce each other throughout the supply chain because of causal relationships within the supply chain.

The thesis is aimed at developing a concept of what constitutes supply chain integration in the building industry. The objective is achieved by comparative analysis of multiple constituting theories, examples from selected manufacturing industries, and cases along the building supply chain. The research is characterised as a descriptive-explorative hypothesis-generating research. The resulting hypotheses indicate the contributions the thesis makes to existing theory rather than developing a new theory. The contributions are particularly relevant for the improved fit of existing theory to the particular combination of characteristics of the building industry, and vice versa.

The development of the concept of supply chain integration has been approached as a theory building exercise using case studies, following Eisenhardt's method (Eisenhardt 1989). The resulting concept is to contribute theoretically to the explanation of building and the understanding of what supply chain integration could and should mean to building. The thesis has used a framework based on building literature and generic theories, and from those inducing an analytical model of supply chain integration. This model has been used for the empirical analysis of the selected applications of supply chain integration in manufacturing and building. Next, the analysis has led to the shaping of the hypotheses, and thus inductively building the theoretical contribution of the thesis in a descriptive and explorative manner.

The concept of supply chain integration has been framed first on the basis of previous thoughts and concepts that have arisen in building practice and literature. This has led to a set of a priori constructs framing the further development of the concept of supply chain integration. The constructs have included indications of lessons the building industry could draw from manufacturing, previous thoughts pointing in the direction of supply chain integration, such as the ideas of the quasi-firm and the systems integrator. In addition, topics that have emerged in recent decades in building literature related to supply chain integration, such as lean construction and partnering, offer further direction to the concept. Although the issues presented have provided initial grounding for the concept of supply chain integration, they have also indicated gaps in the current understanding of supply chain integration in building, which the concept developed has attempted to address.

In addition to the thoughts and topics existing in building practice and literature, the concept of supply chain integration has been framed further within a multiple theoretical framework based on general theory. The framework has aimed to theoretically explain and understand the organisation and functioning of supply chains from four theoretical perspectives of governance: economic governance, production management, inter-firm governance, and social governance. The literature study of the four theoretical perspectives has shown that no single theory appears sufficiently comprehensive to fully explain the multi-faceted phenomenon of the building supply chain. None of the theories as such has represented a complete and useful conceptualisation for supply chain integration in building. Instead, the theories have shown gaps and limitations in their application, particularly related to the characteristics of building. The generic constructs concealed within the theories presented have provided ingredients and further grounding for the development of the concept of supply chain integration in building.

The a priori constructs identified above have been synthesised into an analytical model for supply chain integration. The model represents an operationalisation of the literature and theories explored into a framework of ten factors of analysis initially constituting the conceptualisation of supply chain integration: integration of business activities, collaboration and sourcing strategies, integration of processes and operation, planning and logistics, information sharing, quality management, product development and design, market approach and marketing, cultural alignment, and human resource management. In addition, four development levels of supply chain integration have been identified and added to the factors of analysis. The resulting analytical model has been used as a framework for the empirical analysis and as a basis for the further theory development.

In the empirical part of the thesis, first, supply chain integration practices as applied in six different manufacturing industries have been studied: automotive, aerospace, computers, electronics, clothing and grocery. The analytical model has been used to identify and explore the principles of supply chain integration as applied in the six industries. The explorations have included observations of the organisational and managerial aspects of the extended delivery process throughout the supply chain. The findings from the explorations of the industries, and the example firms per industry, have shown considerable variation in the presence and appearance of supply chain integration practices. The explorations and the examples have been indicative for the understanding and the advancement of supply chain integration broadly across manufacturing industries. The findings indicate generally advanced forms and high levels of supply chain integration. In all cases, the focal firm has shown considerable dominance of the supply chain and an urge to excel operationally with supply chain firms collaboratively. Supply chain firms have been found to be highly committed and aligned to the focal firm and the supply chain.

The building cases have included five firm types in the building industry applying supply chain integration to their parts of the supply chain: clients, developers, designers, builders and suppliers. In all cases, there has been an increase in the level of integration and repetition of the product, process or inter-firm collaboration in the supply chain. The applications of supply chain integration have appeared to vary from less advanced and partial applications to more advanced and complete applications. This variance among the cases could be clarified based on the different businesses, positions and capabilities of the respective firms observed in the cases, in relation to the other firms in the supply chain. Most cases showed an inclination towards increased levels of integration of business functions while other supply chain activities were left relatively untouched, such as integrated information and quality management. The findings of the building case studies have demonstrated relatively less advanced, less comprehensive, quite pragmatic and rather informal approaches to supply chain integration, as compared to theory and manufacturing.

The within and cross case analyses of the manufacturing studies and the building case studies have enabled to explain and understand supply chain integration applications in building and manufacturing, and the similarities and differences between both. Similarities have mainly been found in the reliance on long-term relations with external supply chain partners, joint planning and logistics, and installation of joint teams between firms. However, the level and scope of these issues have shown to be more restricted in building as compared to manufacturing. The factors of analysis have shown to be of equally broad importance in the manufacturing cases, in contrast to the building cases. Particularly important issues in manufacturing have shown to be addressed to a lesser degree in building, such as integrated market approaches, product development and cultural alignment. Typically, integrated approaches to information management and quality management have shown to be underrepresented in building, which are regarded as cornerstones of supply chain integration both in theory and in manufacturing paradoxically.

The applications in manufacturing have appeared to be rather comprehensive and advanced, and apparently following mainstream theory in the area of supply chains. In contrast, the applications in the building cases have seemed rather industry-specific, including pragmatic translations of manufacturing practices adapted to the situation in building practice. The scope and level of the applications in the building cases have varied depending on which firm type represents the focal firm. This situation has appeared to be relatively ambiguous in contrast to manufacturing where it is clear that the lead manufacturer is generally the focal firm. Hierarchical relations and roles between firms in the manufacturing supply chain have shown to be generally clearer. The explanation for these differences could be found in the higher level of maturity and dissemination of the phenomenon of supply chain integration across manufacturing in contrast to building. The understanding and explanation of the aspects of supply chain integration in manufacturing have appeared to be quite unambiguous. In particular, the aspect and level of centralisation exercised by the focal firm is generally undisputed in manufacturing in contrast to building. Additionally, the level of repetitiveness is typically high in manufacturing, extending to virtually all aspects of products, processes, firm structures and inter-firm collaborations.

Based on the confrontation of theory with empirical findings, the theory-building process of the thesis has been finalised by shaping four hypotheses. The hypotheses have represented the concept of supply chain integration as developed in the thesis, respectively reflecting four relations between reduced clusters of factors on the one hand and corresponding reduced clusters of goals on the other. The hypotheses have shown connections with existing theory and supply chain integration applications in manufacturing, and suggest that the building industry could change in this direction as well. Reversely, for improved applicability and operationalisation of the hypotheses to the context of building, their adaptation to that particular context would be helpful, specifically addressing the industry's unique combination of characteristics.

The first hypothesis has implied a relationship between the increased integration and the increased productivity of the supply chain. The relationship has suggested that higher levels of integration of the extended delivery process through the supply chain – including the connections between respective business processes of firms involved – leads to increased effectiveness and efficiency throughout the supply chain. The second hypothesis has implied a relationship between the improved connection to the end market and the improved market position of the supply chain. The relationship has suggested that the improvement of the connection and responsiveness of the integrated supply chain to the end market – supported by increased control of collective market activities and the integrated sales channel towards the end market by the focal firm – leads to improved commercial success of products delivered to the end market, and higher competitiveness of the supply chain and the focal firm in particular as a result. The third hypothesis has implied a relationship between increased collaboration and improved team performance. The relationship has suggested that the intensified alignment of the contributions of firms and individuals to the supply chain – and structuring their collective activities as an integrated team – improves the joint performance as well as the individual performances. The fourth hypothesis has implied a relationship between total quality management and unconditional product delivery. The relationship has suggested that the implementing of total quality management to the integrated delivery process throughout the supply chain – including a commitment of firms to continuous improvement of quality of components, manufacturability of integrated products and quality of processes – leads to the delivery of end products with unconditional quality to the end customer.

The four hypotheses have represented partial descriptions of what constitutes a concept of supply chain integration in building, and collectively have formed the theoretical contribution as it emerged from this thesis. In general, this emergent theory would imply a redefinition of organisational and managerial arrangements between firms in the building supply chain towards higher levels of repetitiveness and integration of products, business processes and inter-firm relations. This would propose the supply chain being governed as an extended enterprise as it were a single firm persisting to exist, beyond the scope of separate projects. Governance mechanisms would then be more centralised into an integrated structure and a single hierarchy. The centralised control would lead to higher levels of synchronisation, knowledge transfer and systematic feedback among firms aimed at value maximisation and market responsiveness, and improved joint performance as a result. The integrated supply chain would be led by a focal firm acting as a supply chain integrator. Alternatively this role could be divided in two partial roles, the demand integrator and the supply integrator, for instance, represented by a client, or respectively by a developer, architect, builder or a group of suppliers. Depending on the firm type being the focal firm, additional cooperative roles could be defined for the other firms in the supply chain, as part of the collective whole of the integrated supply chain.

Samenvatting

(summary in Dutch)

Ketenintegratie in de bouw: de opkomst van geïntegreerde en projectongebonden strategieën in een gefragmenteerde en projectgedreven sector

De bouw is vaak gekritiseerd vanwege de vele verspillingen, het weinige innovatieve gehalte en de lage productiviteit. In die kritieken worden vaak andere sectoren aangehaald, met name de industrie, die producten op een effectievere en efficiëntere wijze produceert dan de bouw blijkbaar. Een van de grote verschillen tussen de bouw en andere sectoren zoals de industrie is onder andere de organisatie en aansturing van de keten. De organisatie en aansturing van de keten heeft een andere vorm en invulling in de bouw dan in de industrie als gevolg van de specifieke verschillende karakteristieken van beide sectoren. In de bouw worden organisatievormen in de keten met name beïnvloed door de eenmalige en tijdelijke aard van projecten, de grote aantallen vaak relatief kleine partijen die betrokken zijn in het initiatief, ontwerp, toelevering en uitvoering, de verspreide macht en controle, en de initiërende rol van opdrachtgevers. De hoge mate van fragmentatie en de lage herhalingsgraad in de bouw hebben ertoe geleid dat zich veel problemen en onvolkomenheden voordoen in het algehele productieapparaat van de bouw. Deze twee fenomenen in de bouwketen van hoge fragmentatie en lage herhalingsgraad hebben geleid tot typerende problemen zoals gebrek aan coördinatie en lage prestaties, die elkaar bovendien lijken te versterken door de keten heen als gevolg van de causale verbanden in de keten.

Het proefschrift is gericht op de ontwikkeling van een concept voor ketenintegratie in de bouw. De doelstelling wordt bereikt door vergelijkend onderzoek tussen verschillende ondersteunende theorieën, voorbeelden uit andere sectoren, en case studies in de bouw. Het onderzoek kan gekarakteriseerd worden als beschrijvend-exploratief hypothese genererend onderzoek. De resulterende hypotheses geven de bijdrage weer die het proefschrift levert aan bestaande theorie in plaats van het ontwikkelen van nieuwe theorie. De bijdrage is met name relevant als verbeterde aansluiting van bestaande theorie op de specifieke combinatie van eigenschappen van de bouw, en vice versa.

De ontwikkeling van het concept voor ketenintegratie wordt benaderd als het bouwen van een theorie met behulp van case studies gebaseerd op de methode van Eisenhardt (1989). Het uiteindelijke concept moet een theoretische bijdrage leveren aan de begripsvorming van de bouwketen en wat ketenintegratie zou kunnen of moeten betekenen voor de bouw. Het onderzoek maakt gebruik van een raamwerk van bouwliteratuur en algemene theorieën, van waaruit een analysemodel voor ketenintegratie wordt afgeleid. Het model is gebruikt voor de empirische analyse van de ketenintegratietoepassingen die zijn onderzocht in de andere sectoren en de bouw. Vervolgens heeft de confrontatie van deze analyses met bestaande theorie geleid tot de vorming van de hypotheses die de theoretische bijdrage van het proefschrift vertegenwoordigen.

Het concept voor ketenintegratie is in eerst instantie verkend op basis van voorgaande gedachten en concepten die zijn ontstaan in de bouwpraktijk en –literatuur. Dit heeft geleid tot een verzameling a priori constructen die de verdere conceptontwikkeling zullen inkaderen. De constructen betreffen in eerst plaats lessen vanuit andere sectoren die voorheen genoemd zijn, voorgaande gedachten in de richting van ketenintegratie zoals het quasi-bedrijf en de systeemintegrator. Daarnaast zijn de afgelopen tientallen jaren onderwerpen voorbij gekomen in de bouwliteratuur die gerelateerd zijn aan ketenintegratie en daar verder richting aan kunnen geven, zoals lean bouwen en partnering. Ondanks de conceptuele inbedding die deze gedachten en onderwerpen bieden aan het concept van ketenintegratie, zijn er ook witte vlekken die zij vertonen in de begripsvorming van ketenintegratie in de bouw, die geadresseerd moeten worden.

Naast de bestaande onderwerpen in de bouwpraktijk en -literatuur, wordt het concept van ketenintegratie verder ingekaderd door een raamwerk van verschillende algemene theorieën. Dat raamwerk is gericht op vergroting van de begripsvorming van de organisatie en aansturing van de keten vanuit vier theoretische gezichtspunten: economische sturing, productiesturing, organisatorische sturing en sociale sturing. De literatuurstudie in deze vier richtingen geeft aan dat geen van de onderzochte theorieën voldoende omvattend is om de bouwketen als fenomeen met diverse verschillende facetten volledig te bevatten. Geen van de theorieën vertegenwoordigt een afdoende complete en bruikbare conceptualisatie van ketenintegratie in de bouw. Integendeel, de theorieën vertonen witte vlekken en beperkingen in hun toepasselijkheid met name in relatie tot de karakteristieken van de bouw. De algemene gedachten die liggen verborgen in de theorieën leveren wel ingrediënten en bieden verdere inbedding voor de conceptontwikkeling van ketenintegratie.

De a priori constructen vanuit de bouwliteratuur en de algemene theorieën zijn samengevoegd en omgevormd tot een analysemodel voor ketenintegratie. Het analysemodel vertegenwoordigt een operationalisering van de literatuur en de theorieën in een raamwerk dat de initiële basis legt voor het concept voor ketenintegratie opgebouwd uit tien analysefactoren: integratie van primaire activiteiten, samenwerking en uitbestedingsstrategieën, integratie van processen en operationele activiteiten, planning en logistiek, informatiedeling, kwaliteitsmanagement, productontwikkeling en ontwerp, marktbenadering en marketing, in lijn brengen van cultuur, en personeelsmanagement. Daarnaast zijn vier toenemende ontwikkelniveaus van ketenintegratie benoemd, voor elk van de analysefactoren. Het resulterende analysemodel wordt toegepast als raamwerk voor de empirische analyse en dient vervolgens als basis voor de verder theorievorming.

Het empirische deel van het onderzoek is gestart met een analyse van de toepassing van ketenintegratie in zes verschillende sectoren: automobiel, vliegtuig, computers, elektronica, kleding en voedingswaren. Het analysemodel is gebruikt als middel voor de identificatie en verkenning van de ketenintegratietoepassingen in de zes sectoren. De verkenning is gericht op de verschillende aspecten van de organisatie en aansturing van het integrale leveringsproces door de keten heen. De bevindingen van de verkenningen van de verschillende sectoren, en de bijbehorende voorbeeldbedrijven per sector, laten een gevarieerd beeld zien van de aanwezigheid en verschijningsvorm van ketenintegratie. De verkenningen en voorbeelden zijn indicatief geweest voor de begripsvorming en het niveau van ketenintegratie in industriële sectoren. De bevindingen geven een algemeen beeld weer van vergevorderde vormen en hoge niveaus van ketenintegratie. In alle gevallen bleek de centrale ketenspeler, de 'focal firm', dominant in de keten te zijn en de drang te hebben te willen excelleren samen met de andere partijen in de keten. De andere ketenpartijen bleken in hoge mate gecommitteerd en gekoppeld aan de focal firm en de aan de keten.

De negen case studies in de bouw waren verspreid over vijf typen van partijen in de bouwketen die een vorm van ketenintegratie toepasten op hun deel van de keten en soms in samenwerking met andere partijen: opdrachtgevers, ontwikkelaars, ontwerpers, bouwers en leveranciers. In alle gevallen was er sprake van een verhoogd niveau van integratie en herhaalbaarheid van product, proces en organisatievorm in de keten. De ketenintegratietoepassingen bleken echter te variëren van minder vergevorderde deeltoepassingen tot vergevorderde integrale toepassingen. Deze variatie kan verklaard worden op basis van de verschillende business, positie en mogelijkheden die verschillende partijen in de cases hebben in relatie tot de andere partijen in de keten. De meeste cases lieten een neiging zien naar integratie van partijen van primaire activiteiten binnen de eigen organisatie terwijl externe activiteiten relatief onberoerd bleven, zoals integraal informatiemanagement en kwaliteitsmanagement. De bevindingen uit de case studies in de bouw lieten in het algemeen relatief minder vergevorderde, minder veelomvattende, nogal pragmatische en informele benaderingen van ketenintegratie zijn, in vergelijking met de theorie en de andere sectoren.

De analyses van en tussen de case studies in de bouw en de studies in de andere sectoren hebben bijgedragen aan de verklaring en de begripsvorming van de ketenintegratietoepassingen in de bouw en de andere sectoren, alsmede de overeenkomsten en verschillen daartussen. Overeenkomsten bleken met name te bestaan in de lange termijn relaties die werden aangegaan met ketenpartners, gezamenlijke planning en logistiek en de vorming van gezamenlijke teams tussen bedrijven. Echter het niveau en de omvang van deze toepassingen waren in de bouw minder verstrekkend dan in de andere sectoren. De analysefactoren van het analysemodel bleken in het geval van de andere sectoren van evenredig groot belang in tegenstelling tot de bouw. Met name belangrijke zaken in de andere sectoren bleken in de bouwketen van minder belang, zoals integrale marktbenaderingen, integrale productontwikkeling en in lijn brengen van bedrijfsculturen. In het bijzonder bleken integrale benaderingen van informatiemanagement en kwaliteitsmanagement ondervertegenwoordigd in de bouw, terwijl die zaken zowel in de theorie als in de andere sectoren paradoxaal genoeg als pijlers van ketenintegratie worden beschouwd.

De ketenintegratietoepassingen in de andere sectoren bleken over het algemeen breed en vergevorderd, en in lijn met de algemene theorie. De toepassingen in de bouw bleken daarentegen relatief specifiek te zijn voor de bouw, en vaak neer te komen op pragmatische vertalingen en aanpassingen van industriële toepassingen aan de situatie in de bouwpraktijk. De omvang en het niveau van de toepassingen in de bouw bleken te variëren naar gelang welk type partij de focal firm representeerde. Deze situatie was veel minder eenduidig dan de in de andere sectoren waar het duidelijk is dat de hoofdproducent doorgaans de focal firm is. De hiërarchische verhoudingen tussen partijen in de keten is doorgaans ook duidelijker in de andere sectoren. De verklaring voor deze verschillen kan mede worden gevonden in de grotere volwassenheid, wijdere verspreiding en formelere status van het fenomeen ketenintegratie in andere sectoren in tegenstelling tot de bouw. De begripsvorming en uitleg die wordt toegedicht aan aspecten van ketenintegratie is navenant eenduidig in andere sectoren. Met name het aspect en niveau van centralisatie en de geprogrammeerde centrale coördinatie die wordt uitgeoefend door de focal firm staat in het algemeen niet ter discussie in andere sectoren in tegenstelling tot de bouw. Daarnaast is het herhalingsgraad doorgaans hoog in de andere sectoren, en beslaat nagenoeg alle aspecten van producten, deelproducten, processen, organisatievormen en samenwerking.

Tenslotte is op basis van de confrontatie tussen theorie en praktijk zoals gevonden door het onderzoek heen, het proces van theorie bouwen afgerond met de vorming van vier hypotheses. De hypotheses vertegenwoordigen het ontwikkelde concept van ketenintegratie, en representeren op gecondenseerde wijze de relaties tussen gereduceerde groepen van factoren aan de ene kant, en bijbehorende gereduceerde groepen van doelen van ketenintegratie aan de andere kant. De hypotheses blijken relaties te hebben met bestaande theorieën en ketenintegratietoepassingen in andere sectoren, en suggereren dat de bouw in deze richtingen zou kunnen veranderen door toepassing van ketenintegratie. Andersom zou het de toepasselijkheid en operationaliseerbaarheid van de hypotheses, en dus van het concept van ketenintegratie, ten goede komen als zij verder gespecificeerd worden, in de context van de unieke combinatie van karakteristieken van de bouw.

De eerste hypothese impliceert een relatie tussen de toename van integratie en toename van productiviteit in de keten. Deze relatie suggereert dat een hoger niveau van integratie van het integrale leveringsproces door de keten heen – en daarmee de koppelingen tussen de respectievelijke primaire processen van betrokken partijen – leidt tot hogere effectiviteit en efficiëntie door de keten heen. De tweede hypothese impliceert een relatie tussen een betere koppeling met de eindmarkt en een betere marktpositie van de keten. Deze relatie suggereert dat een verbetering van de verbinding en reactiesnelheid van de geïntegreerde keten tot de eindmarkt – gefaciliteerd door een hogere mate van coördinatie door de focal firm van de marktactiviteiten en het verkoopkanaal van de keten naar de eindmarkt – leidt tot een beter commercieel succes van producten die geleverd worden aan de eindmarkt, en als gevolg een betere concurrentiepositie van de keten en de focal firm in het bijzonder. De derde hypothese impliceert een relatie tussen meer samenwerking en verbeterde teamprestatie. Deze relatie suggereert dat het actief op een lijn brengen van de bijdragen van partijen en individuen aan de keten – en het structureren van de verzameling van hun activiteiten als een geïntegreerd team – leidt tot een betere gezamenlijke prestatie alsmede betere individuele prestaties. De vierde hypothese impliceert een relatie tussen integraal kwaliteitsmanagement en onconditionele productlevering. Deze relatie suggereert dat de invoering van integraal kwaliteitsmanagement binnen het integrale leveringsproces door de hele keten heen – en het commitment van partijen in de keten aan continue kwaliteitsverbetering van deelproducten, maakbaarheid van het eindproduct en proceskwaliteit – leidt tot levering van eindproducten van onconditionele kwaliteit aan de eindklant.

De vier hypotheses vertegenwoordigen deelbeschrijvingen van wat een concept voor ketenintegratie in de bouw zou inhouden. Gezamenlijk vormen de hypotheses de theoretische bijdrage zoals die uit dit proefschrift naar voren is gekomen. In het algemeen stelt de theoretische bijdrage een herdefiniëring van de organisatie en aansturing van partijen in de bouwketen in de richting van meer integratie en hogere herhalingsgraad van producten, processen en relaties tussen partijen. Dit zou leiden tot de situatie dat de keten zou worden bestuurd als een 'extended enterprise', al ware het een enkel bedrijf dat blijft bestaan over projecten heen. Besturingsmechanismen zouden dan meer gecentraliseerd worden binnen een geïntegreerde structuur en een eenduidige hiërarchie in de keten. De centrale coördinatie zou leiden tot een hogere mate van synchronisatie, kennisdeling en systematische feedback tussen partijen gericht op maximalisatie van de waardecreatie in de keten en reactiesnelheid richting eindklant, met een verbeterd gezamenlijk operationeel en commercieel resultaat tot gevolg. De geïntegreerde keten zou geleid worden door de focal firm die optreedt als ketenintegrator. Als alternatief zou deze rol kunnen worden opgedeeld in twee rollen: de vraagintegrator en de aanbodintegrator, bijvoorbeeld, ingevuld door een opdrachtgever, of respectievelijk door een ontwikkelaar, ontwerper, bouwer of een groep leveranciers. Afhankelijk van welke partij de focal firm zou zijn, zouden aanvullende samenwerkingsrollen worden gedefinieerd voor de andere partijen in de keten, als deel van het collectief dat de geïntegreerde keten vormt.

Curriculum Vitae

Ruben Vrijhoef was born on the 26th of December 1972 in Wormerveer in the Netherlands. He went to secondary school, i.e. preparatory university education in the period from 1985 till 1991, followed by university education in the period from 1991 till 1998 at the Delft University of Technology. As part of his master study, he worked as a junior researcher from 1997 till 1998 at the VTT Technical Research Centre of Finland. In June 1998 he graduated as a civil engineer from the Delft University of Technology. From 1998 till 2007, he worked as a research consultant at TNO Built Environment and Geosciences in the field of building process innovation. In addition, from 1998 till 2000, he worked at HBG construction company (currently Royal BAM). Since 2002, he has been an active member of the Centre for Process Innovation in Building and Construction at the Delft University of Technology. From 2002 till 2004, he acted as a programme manager to the national innovation programme for the Dutch building industry PSIBouw. In 2005, he was a visiting researcher at the Salford Centre for Research and Innovation (SCRI) of the University of Salford in the UK. From 2006 till 2009, he was a part-time freelance writer and editor to a number of trade journals for the Dutch building sector. From March 2007 till February 2010, he joined the Design and Construction Processes research group at the Faculty of Civil Engineering of the Delft University of Technology. Since February 2010, he has moved to the Design and Construction Management research group at the Faculty of Architecture. His professional activities include both fundamental and applied research of process innovations in building, publication of scientific and industry papers, supervision of master students, and advising and guiding industry partners applying supply chain integration and additional process innovations in practice. His research interests include construction management, process innovation, supply chain management, lean construction, and comparative studies between sectors of industry. He has been a member of the CIB network and the International Group for Lean Construction (IGLC). He has published numerous scientific and industry papers, and won a few best paper awards.

Publications

Vrijhoef, R. and De Ridder, H.A.J. (2009). "A systems approach to construction supply chain integration". In: Atkin, B. and Borgbrant, J. (eds.). *Performance Improvement in Construction Management* (pp. 34-44). Spon Press, London.

Vrijhoef, R. and London, K.A. (2009). "Review of organizational approaches to the construction supply chain". In: O'brien, W.J., Formoso, C.T., Vrijhoef, R. and London, K.A. (eds.). *Construction Supply Chain Management Handbook*, pp. 10/1-10/19. CRC Press, Boca Raton.

Vrijhoef, R. and De Ridder, H.A.J. (2008). "Methodological challenges of knowledge production in construction management research: the case of a postgraduate research into supply chain integration". In: Dainty, A. (ed.). *Proceedings of the 24th Annual ARCOM Conference*, pp. 707-716. Association of Researchers in Construction Management, Reading.

Vrijhoef, R. and De Ridder, H.A.J. (2008). "Mode 2 knowledge production in construction management research: the case of a postgraduate research into supply chain integration". In: Cásenský, M., Ahmed, V., Eaton, D. and Sutrisna, M. (eds.). *Proceedings of the BuHu 8th International Postgraduate Research Conference 2008*, pp. 374-381. University of Salford, Salford.

Vrijhoef, R. and De Ridder, H.A.J. (2007). "A systems approach for developing a model of construction supply chain integration". In: Borgbrant, J. and Atkin, B. (eds.). *Proceedings 4th Nordic Conference on Construction Economics and Organisation*, pp. 1-11. Lulea University of Technology, Lulea.

Vrijhoef, R. and De Ridder, H.A.J. (2007). "A systems approach to supply chain integration in construction". In: Hughes, W. (ed.). *Proceedings CME25 Conference*, pp. 1-12. University of Reading, Reading.

Vrijhoef, R. and De Ridder, H.A.J. (2007). "Integrating the construction supply chain by applying systems thinking". In: Egbu, C.O. and Tong, M.K.L. (eds.). *Proceedings of the Third Scottish Conference for Postgraduate Researchers of the Built and Natural Environment*, pp. 469-479. Glasgow Caledonian University, Glasgow.

Vrijhoef, R. and De Ridder, H.A.J. (2007). "Supply chain systems engineering in construction". In: s.n. (ed.). *Proceedings CIB World Building Congress*, pp. 279-291. CIB, Cape Town.

Vrijhoef, R. and De Ridder, H.A.J. (2006). "Systems thinking approach to supply chain integration in construction". In: Pietroforte, R., De Angelis, E. and Polverino, F. (eds.).

Proceedings Construction in the XXI Century: Local and Global Challenge, pp. 1-11. Edizione Scientifiche Italiane, Napoli.

Vrijhoef, R. and Koskela, L. (2005). "A critical review of construction as a project-based industry: identifying paths towards a project-independent approach to construction". In: Kazi, A.S. (ed.). *Proceedings Learning from Experience: New Challenges, Theories and Practices in Construction,* pp. 13-24. VTT, Espoo Finland.

Vrijhoef, R. and Koskela, L. (2005). "Revisiting the three peculiarities of production in construction". In: s.n. (ed.). *Proceedings 13th International Group for Lean Construction Conference*, pp. 19-27. Unitec, New Zealand.

Vrijhoef, R. and Koskela, L. (2005). "Structural and contextual comparison of construction to other project-based industries". In: Ruddock, L. et al. (eds.). *Proceedings 5th International Postgraduate Research Conference April 14th-15th 2005,* pp. 537-548. University of Salford, Salford.

Vrijhoef, R. and De Ridder, H.A.J. (2005). "Supply chain integration for achieving best value for construction clients: client-driven versus supplier-driven integration". In: Sidwell, A.C. (ed.). *Proceedings QUT Research Week 2005*, pp. 260-271. Queensland University of Technology, Brisbane.

Vrijhoef, R. and Voordijk, H. (2004). "Improving supply chain management in construction: what can be learned from the electronics industry". In: s.n. (ed.). *CIB World Building Congress 2004: Building for the Future,* pp. 1-11. CIB, Toronto.

Vrijhoef, R. and Voordijk, H. (2004). "Supply chain implications of platform strategies in construction". In: Ruddock, L. et al. (eds.). *Proceedings of the Fourth International Postgraduate Research in the Built and Human Environment Conference,* volume 1, pp. 77-87. Blackwell Publishing, Manchester.

Vrijhoef, R. and Tong, K.L. (2004). "Understanding construction supply chains: an adaptive network approach". In: L. Ruddock et al. (eds.). *Proceedings of the Fourth International Postgraduate Research in the Built and Human Environment Conference,* volume 2, pp. 658-669. Blackwell Publishing, Manchester.

Vrijhoef, R. and Tong, M.K.L. (2004). "Understanding constructions as a complex and dynamic system: an adaptive network approach". In: Bertelsen, S. and Formoso, C.T. (eds.). *Proceedings 12th Annual Conference on Lean Construction,* pp. 61-73. Helsingor, Denmark.

Vrijhoef, R. and Voordijk, H. (2003). "Improving supply chain management in construction: what can be learned from the aerospace industry". In: Greenwood, D. (ed.). *Proceedings ARCOM 19th Annual Conference,* volume 2, pp. 837-846. ARCOM, Reading.

Vrijhoef, R. and Voordijk, H. (2003). "Improving supply chain management in construction: what can be learned from the clothing and textile industry". In: Leung, B.Y.P. et al. (eds.). *Proceedings Conference on Innovation in Construction and Real Estate*, pp. 89-100. The Hong Kong Polytechnic University, Hong Kong.

Vrijhoef, R. and Voordijk, H. (2003). "Improving supply chain management in construction: what can be learned from the food and grocery sector". In: Ofori, G. and Ling, F.Y.Y. (eds.). *Proceedings Conference Knowledge Construction*, volume 2, pp. 155-165. National University of Singapore, Singapore.

Vrijhoef, R., Koskela, L. and Voordijk, H. (2003). "Understanding construction supply chains: a multiple theoretical approach to inter-organizational relationship in construction". In: Martinez, J.C. and Formoso, C.T. (eds.). *Proceedings 11th Annual Conference on Lean Construction*, pp. 280-292. Virginia Polytechnic institute and State University, Blacksburg.

Vrijhoef, R., Cuperus, Y.J. and Voordijk, H. (2002). "Exploring the connection between open building and lean construction: defining a postponement strategy for supply chain management". In: Formoso, C.T. and Ballard, G. (eds.). *Proceedings 10th Annual Conference on Lean Construction*, pp. 149-160. Federal University of Rio Grande do Sul, Gramado.

Vrijhoef, R., Koskela, L. and Howell, G. (2001). "Understanding construction supply chains: an alternative interpretation". In: *Proceedings 9th Annual Lean Construction Conference*. National University Singapore, Singapore.

Vrijhoef, R. and Koskela, L. (2000). "The four roles of supply chain management in construction". *European Journal of Purchasing & Supply Management, 6*(3-4), 169-178.

Vrijhoef, R. and Koskela, L. (1999). "Roles of supply chain management in construction". In: Tommelein, I. (ed.). *Proceedings 7th Annual Lean Construction Conference*, pp. 133-146. University of California, Berkeley.